AKADEMIE DER WISSENSCHAFTEN DER DDR

Veröffentlichungen des Forschungsbereichs Geo- und Kosmoswissenschaften

Herausgegeben von H.-J. Treder

Heft 9

Commission for Multilateral Cooperation of Academies of Sciences
of Socialist Countries on the Complex Problem
„Planetary Geophysical Studies"
CAPG-Working Group 1.11:
„Physical Properties of Rocks and Minerals
at High Thermodynamic Parameters"

Theoretical and Experimental Investigations of Physical Properties of Rocks and Minerals under Extreme *p*, *T*-conditions

edited by H. STILLER, M. P. VOLAROVICH
associate editors T. S. LEBEDEV, H. VOLLSTÄDT
executive editor D. FREUND

with 146 Figures

AKADEMIE-VERLAG · BERLIN
1979

Erschienen im Akademie-Verlag, DDR-108 Berlin, Leipziger Straße 3—4

Lizenznummer: 202 · 100/463/79
Umschlaggestaltung: Nina Striewski
Gesamtherstellung: VEB Druckerei „Thomas Müntzer", 582 Bad Langensalza
Bestellnummer: 762 688 7 (2156/9) · LSV 1435
Printed in GDR
DDR 38,— M

Contents

CHAPTER III

CHAPTER IV

CHAPTER V

CHAPTER VI

PREFACE

The present monograph of the working group 1.11 of the Commission of Academies of Sciences of Socialist Countries on the Complex Problem "Planetary Geophysical Studies" represents a summary of the results obtained in the past 5 years in this working group. The monograph is a logical continuation of the working group's first common publication "High pressure and temperature studies of physical properties of rocks and minerals" (appeared by "Naukova Dumka", Kiev, 1977).

The within CAPG co-operated laboratories are engaged since foundation of this working group in 1970, in providing of the theoretical and experimental contributions to behaviour of rocks and minerals at extreme *p*,*T*-conditions.

Owing to appropriate specialization of different laboratories and by continuous intensive co-operation essential results in physical behaviour of materials at conditions of Earth's interior are reached.

These during permanent meetings dicussed and on different international assemblies reported results form the content of this monograph.

The working group 1.11 would like to give with this publication in addition to the representation of new results of common investigations simultaneously a contribution for prospective forming of the geophysical solid investigation within the Geodynamics Project and his connected follow projects. The editors are very indebted to all authors and co-workers of all in this working group acted laboratories for successful co-operation in this paper. Simultaneously we express our thanks to the publishing house for his readiness and reliable designing of the book and to D. Freund for the difficult lesson of manuscript preparation.

H. Stiller
M. P. Volarovich

INTRODUCTION

The present book is divided into various methods for the describing of different physical properties at extreme p,T-conditions, into methodical explanation and theoretical and geophysical consequences.

In chapter 1 experimental data of elastical anisotropy behaviour at extreme pressures are described. In like manner ultrasonic data of various rock-forming minerals and rocks of different regions are important.

The stronger confinement of measurements at defined hydrostatic pressure and the comparison with petrological properties represent an enlargement. Subsequent the international trend and in connection with the present topical problem of earthquake prediction, contributions to the deformation behaviours of rocks under high pressure and temperature become interesting.

In an adequate chapter is made allowance for the importance of electrical behaviour at extreme conditions of rocks, minerals and synthetical substances relevant to the Earth.

With regard to investigations of Earth's crust rocks and to interpretation of results, interesting in composition of lithosphere, magnetic and thermical properties are very sensitive characteristic data. Some new methods for the determination of thermical properties and magnetic behaviour are described in a special chapter. Lunar rocks investigations follow.

The interpretation of different physical properties of rocks and minerals at extreme p,T-conditions for the complicated general behaviour of the Earth is an important task of the working group 1.11.

Analògously to this importance some articles deal with geophysical conclusions and applications of experimental data. To this belong theoretical works describing as well mathematical models for calculation of phase transformations as the influence of pressure and temperature on the elastic behaviour of rocks and model substances.

Finally an important paper discusses phenomenological models of wave propagation in porous and cracked rocks in connection with the earthquake mechanism.

CHAPTER I

Elastic Properties

APPARATUS FOR TEMPERATURE STUDIES OF THE MINERAL MATTER ELASTIC PROPERTIES UNDER HIGH HYDROSTATIC PRESSURES

T. S. Lebedev, V. A. Korchin, V. T. Vasilyaka, V. I. Shapoval

In experimental studies of physical properties of rocks and minerals under conditions of the simultaneous effect of high pressures and temperatures various heaters are utilized. Their design is dependent, in the first place, of dimensions of high pressure apparatus, volumes of heated matter, possible inhomogeneity of the temperature distribution in the samples under study, duration of the experiment and power of heating cells [1—9, 17—21, 23, 24]. Heaters may be placed inside the high pressure chamber (inside heaters) or at the surface (outside heaters).

In the hydrostatic pressure apparatus the outside heating cells of wire or foil can be utilized. The exploitation of these heaters is most suitable in experiments when the change of the mineral matter physical properties (elastic, thermal, electric, magnetic, etc.) is being studied. The construction of these devices varies with the working chamber dimensions and the goal of the experiment. However, the main, listed below, requirements to are common for all the heaters used in hydrostatic pressure apparatus irrespective of its dimensions and conditions of experiments. The heating cell should be of sufficient power with minimum of energy dissipation into the inner walls of pressure vessel. A good quality of closing device and obturator thermoisolation from the heater as well as minimum dimensions of the latter are also necessary.

What is additionally required form the heaters used in the high *PT* experiments. It is important to obtain the least temperature gradient over the generatrix and radius of the sample. The rigid connection of the heater with the closing locker plug of high pressure apparatus should be provided as well as easiness of inserting the sample under study into heater. Besides, it is important enough to provide a high thermoisolation of ultrasonic detectors.

The construction of such heaters have facilitated the experiments according to the programs of certain *PT*-regimes of deep interiors modelled in high hydrostatic pressure chambers. The data on the nature and regularities of the velocity change and some other elastic characteristics of rocks and minerals revealed under the simultaneous *PT* effect are also most important in deciphering the field seismic experiment data.

To meet the above methodical and technical requirements a new construction of the high temperature electroheater (fig. 1, 2) is developed by the authors at the department of physical properties of the Earth's matter, Institute of Geophysics, Academy of Sciences of the Ukr. SSR. The heater is placed inside the high hydrostatic pressure chamber [13].

The heater is meant for utilization in studying elastic and some other physical

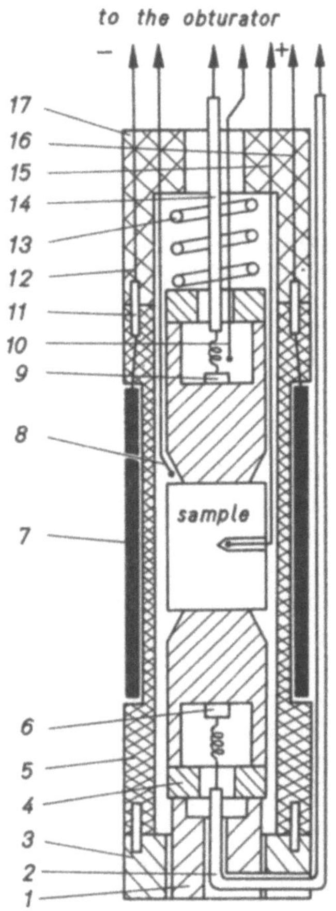

Fig. 1. Inside circuit of hydrostatic pressure apparatus for studies of mineral matter elastic properties under high thermodynamic regimes. 1 — screw support for sample and detector positioning control, 2, 14 — high frequency leadouts, 3 — nut-support, 4 — ultrasonic detector shell, 5 — heater carcass, 6, 9 — piezoceramic detectors, 7 — heating cell, 8, 10, 15 — thermocouples, 11 — heater plug connector, 12, 16 — heating cell lead-ins, 13 — pressure spring, 17 — cantilever.

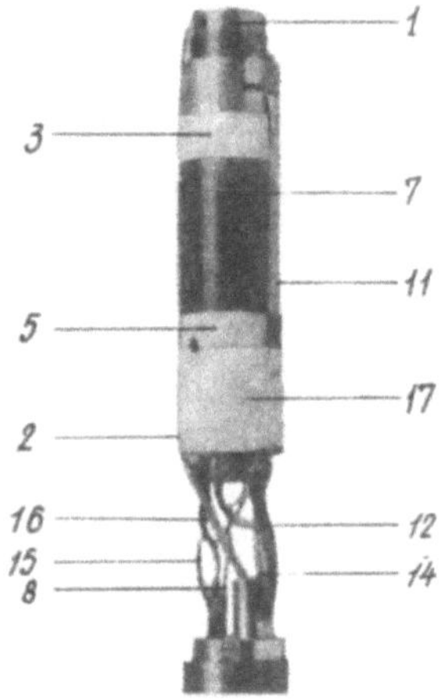

Fig. 2. General view of inner design of hydrostatic pressure apparatus for ultrasonic investigations under high PT regimes. Legend as for fig. 1.

properties of the mineral matter. It is filled into the high hydrostatic pressure apparatus, the inside diameter of working chamber being 32 mm and diameter of the sample under study 22 mm [9].

The main element of the electroheating device is a thin-walled cylindrical carcass of pyrophyllite [5]. Its middle part contains the heating cell in the form of a Ni-Cr-strip put within the spiral groove [7]. From the outside the heater is insulated by the sheet mica and a glass cord (fig. 1, 2).

The heater containing carcass is connected with a bowlshaped supporting cylinder (17) of pyrophillite. On the opposite side, the construction is closed by a brass nut (3) wherein a hollow screw is moving (1). The latter is for fitting the ultrasonic detectors and sample within the heater.

For the first in the practice of such experiments, the parts of the heater carcass (5, 17) are made of pyrophyllite (aluminium hydrous silicate — $Al_2O_3 \cdot 4SiO_2 \cdot H_2O$). Compared with the used conventionally for similar purposes material, the pyrophyllite has some advantages (metals, ceramics, BT-glass, mica). The natural pyrophyllite is processed easily, so the details can be variously shaped can take any shape. Besides, the articles work-pieces of pyrophyllite baked at 1200 °C acquire very high and lasting strength as well as good electric and thermoinsulation qualities. The low termoconductivity of pyrophyllite provides the minimum dissipation of thermal energy over the heater generatrix. The bowl-shaped supporting cylinder (17) serves as a thermal screen to lower the heating of the electric lead-in, high-frequency wires, obturator and lock packing (seal, tightening gasket, gland). By now, the proposed heater has proved to be a success in experiments on elastic properties of rocks performed with the apparatus wherein hydrostatic pressure is 6 kbar and temperature is 350 °C.

As the pressure transmitting medium the polyethilsiloxanic No. 5 liquid or PMS-100 (GOST 13032-67) were used. Heating cell was connected to the BSA-5 d. c. generator. Consumed power of the heater in the maximum temperature domain was 200—250 Wt for current intensity of 8—10 A.

The heating cell was carefully calibrated by varying the supply conditions and temperature with pressure. Since the elastic properties of rocks were investigated under the simultaneous effect of high pressures and temperatures by the programs modelling the corresponding *PT*-conditions of specific geologic regions [8, 12, 16] it was necessary to choose the most reasonable conditions for the heater work to provide an equal duration of experiments irrespective of the regime adopted. The time measurements for the elastic pulse propagation through the sample under various *PT*-conditions is made in about 20 descrete points. After a certain pressure value had been achieved the heater was turned in. The voltage potential on the heater was adjusted according to the $U = f(T)$ or $T = f(U, t)$ curves, where U was supply voltage, T — temperature sample, t — experiment duration. At a certain temperature, the sample should be left alone for 5—10 minutes for regaining the thermal balance within the cell and reducing to a minimum the temperature gradient along the generatrix and radius of the sample.

Special experiments with several thermocouples have shown that the 5 minutes of keeping the sample at 300 °C (heating to this level takes about 90 minutes) provides the average center symmetrical gradient of the temperature change over the sample generatrix is $0.7 \div 0{,}5$ °C/mm and radial gradient is $0.3 \div 0.2$ °C/mm. Thus, with the control provided by two thermocouples the sample temperature may be evaluated within the error of $\pm 2 \div 3$ °C. The gradients may be reduced at the expense of the increase of linear dimensions of the heating cell. However, these are limited by character of design, in the first place, as well as the conditions of work of the ultrasonic detectors inside the heating cell (fig. 1, position *4*, *6*, *9*). The detector construction includes buffer rods to provide a satisfactory acoustical agreement in the measuring circuit (piezoelectric element-buffer-sample) and high resolution of transverse and longitudinal elastic signals [11]. Besides, with the properly chosen material of the buffers, they may be used also as thermal buffers to provide a low heat flow into the piezoceramic element. Fused quartz may be chosen as such a material. As examination has shown the rods as long as 20 mm security a temperature drop of about 60—80 °C. This makes it possible to work with the piezoceramic detectors characterized by the temperature of depolarization of 250 bis 280 °C (achieved on the sample is T of 310—360 °C). During experiments one of the thermocouples is controlling constantly the piezoceramic element temperature. The detector ist glued to the sample by water-glass that provides a good acoustical contact at temperature as high as 300—400 °C.

The high temperature studies in high hydrostatic pressure chambers has necessitated the developing new reliable designs of electric lead-ins that can retain the electro-insulating qualities even at high temperatures. Our design of obturator allows to lead in the pressure

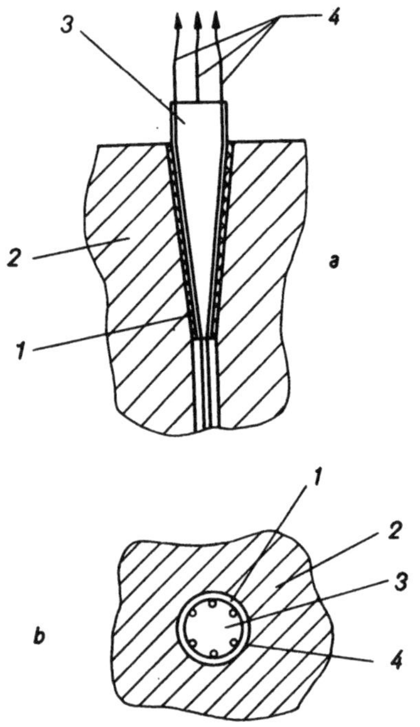

Fig. 3. Design drawing of the cone for continuous thermocouples lead-in into the pressure vessel.
Sections: a — along the cone axis, b — cross section.
1 — nonhardened pyrophyllite cone, 2 — steel plug, 3 — hardened pyrohyllite cone, 4 — thermocouple wire.

chamber three-six thermocouples without breaks and two high-frequency signals (fig. 2, 3). One lead-in is used for the voltage supply to electroheating device. Negative contact of the heating element is connected to the frame of the high pressure apparatus. For the high-frequency signals the conventional lead-ins were used in the forms of steel cones isolated by a glass yarn glued by BF-2. Insertion of thermocouples was provided by a new construction of lead-in used for the first time in the high pressure technique. The main element of the lead-in (fig. 3) is a cone (3) made of pyrophyllite. Along the side surface of the cone thin kerfs were cut for the thermocouple wires. Then the cone was dried and annealed. The thermocouple wires were introduced in the notches of the thermally treated cone — and the top was sealed by the BF-2 glue polymerized by a certain technology. Then the cone (3) is inserted into a hollow cone (1) with the subsequent pressing-in into the conic aperture of the high pressure Apparatus cork. Direct current resistance between the thermocouple and the cork mass near the lead-in is $10^9 \div 10^{10}$ ohm. It remains unchanged for many months of running the high pressure apparatus. The lead-in system developed by the authors has shown a good stability in quite a number of experiments run at the in cell temperature to 320 °C and pressure to 6 kbar. The possibility of admission of several conductors into the high hydrostatic pressure chamber by means of one lead-in has simplified the obturator construction. The obturator has become more compact and production of the apparatus cork less laborous.

The use of the described high temperature cell has facillitated the experiments on investigation of the change of longitudinal and transverse ultrasonic velocities in the mineral matter samples under the thermodynamic regimes corresponding to deep-seated crustal zones [10, 11].

The experiments run under various *PT* conditions have shown that the $V_{p,s}$ behaviour with depth is rather complicated and characterized, as a rule, by the presence of extremum zones. The location of the latter depends on composition of the object under study and specific thermodynamic conditions of the region. Origination of the said zones of minimum and maximum velocity at various depths is due to interaction of the processes thermal change and dynamic consolidation of the rock.

The high hydrostatic pressure and temperature experiments using the programmed modelling technique [15, 16] have also made possible the appraisal of the effect of the pressure distribution inhomogeneity inherent in quasihydrostatic conditions on the velocity change [8, 10].

EQUIPMENT FOR ANISOTROPY INVESTIGATIONS OF ROCK SAMPLES UNDER HIGH PRESSURES

H. J. MÜLLER, S. RAAB, U. SEIPOLD

The knowledge of the elastic properties of rocks and minerals as functions of pressure and temperature is an important means of solving various geophysical and geological problems. For the laboratory measurements of the elastic parameters, generally the various ultrasonic methods have been established. Here it is assumed that the dispersion can be neglected for elastic waves.

For generating the required extreme thermodynamic conditions, the use of hydrostatic pressure chambers is restricted to the lower pressure ranges. From among the great number of quasi-hydrostatic pressure systems, piston-cylinder apparatus are used in most cases for routine operation in the determination of the elastic properties. By means of the quasi-hydrostatic generation of pressure it is possible to achieve higher pressures, but there is a considerable difficulty in that the ultrasonic transmitters and receivers cannot be accommodated in the pressure chamber. In what follows, a six-anvil equipment shall be described, by means of which — using an ultrasonic pulse technique — the propagation velocity of elastic longitudinal waves can be measured simultaneously in three mutually orthogonal directions in a cubic rock sample. Thus it is possible to measure the variation of anisotropy of the elastic properties under pressure. Such a hexahedric equipment has the advantage of a large sample volume and a high maximum pressure. Its disadvantage is the much higher technical effort and the more complicated handling as compared to piston-cylinder arrangements. The forces required for the generation of pressure are provided by three independent hydraulic sets in the triaxial equipment used in this case.

In the vertical direction (Z) a maximum force of 600 Mgf is available, whereas the maximum forces in the two horizontal directions (x and y) are 300 Mgf each.

The six anvils enclose a cubical sample space. They are connected to the hydraulic pistons through a positive supporting system. The anvils are electrically insulated form each other, so that they can be used for the current supply to an internal heating system and for resistance measurements. The upper vertical anvil has a motor-driven lifting and lowering device. This device is useful in the insertion and replacement of the samples. The die systems consist of two parts, a base component (anvil holder) and the actual anvil. The base component carries the anvil in a tapered hole and supports it laterally. On the side of the pressure chamber the anvil end has the shape of a pyramid frustum with a slope angle of 45°, and has a square working face with an edge length of 20 mm. The ultrasonic transducers are arranged at the rear side of the anvil in a recess of the base component. Thus the travelling distance of the ultrasonic pulse is relatively short. Disturbing interfaces, which lead to a signal attenuation, are avoided. The base components and the anvils are made of differently hardened steels. The angle of aperture of the anvil holder is 40°. At this value a force distribution results, which provides a good supporting action for the anvil. In an originally used construction with flatly supported anvils this supporting action was lacking. This frequently led to the formation of cracks which developed from the recesses provided for the ultrasonic transducers and resulted in a anvil fracture.

For a high efficiency of force utilization and a long service life, an exact alignment of the anvils is required. The sealings formed between the anvils should, if possible, have the same thickness. The sample must be placed exactly in the centre of the sample space. For checking the anvil alignment, aluminium alloy cubes having an edge length of 24 mm were pressed and then the anvil impressions were measured by means of a micrometer screw. Using a system of adjusting devices it is possible to adjust the optimum anvil position. By means of repetitive checking followed by correction, the alignment can be improved to a degree where the deviations are less than 0.1 mm.

Fig. 1 shows the sample arrangement for measuring the propagation velocity v_p of elastic longitudinal waves. Pyrophyllite was used as a pressure transmitting medium. The employed material exhibited a considerable anisotropy. Under normal conditions the factor of anisotropy,

$$k = \frac{v_{\max}}{v_{\min}}$$

had a value of 1.3. Therefore the pyrophyllite slabs used for encasing the sample were cut out of the starting material in a well-defined way. In determining the sample size, two contradictory requirements must be taken into account. On the one hand the portion of the total volume occupied by the sample should be as large as possible in order to minimize the effects of the pressure transmitting medium. On the other hand the mechanical stresses occurring in the range of the gap between the anvils must not result in the destruction of the sample. In the pressure range used, it was found that the optimum sample edge length is 18 mm when the sample is encased by a pyrophyllite layer being 3 mm thick, so that the resulting edge length of the sample container is 24 mm. With these dimensions it was even possible to investigate single crystals in the pressure test without being destroyed.

The pressure calibration of the device is performed using the phase transitions of bismuth and thallium. For that purpose wires having a diameter of 0.4 mm were made by extrusion from these metals. They were inserted into argentic chloride cylinders. Contacts were applied by means of platinum wires and foils, respectively.

A pulse technique is used for measuring the propagation velocity. It reduces the velocity measurement to measuring a short time interval. For determining the length variation of the sample, the displacement of the hydraulic pistons is measured by means of inductance pick-offs. The effects of anvil deformation and of the pyrophyllite were investigated by calibration measurements and corrected during the evaluation. The accuracy of measurement is to a high degree determined by the accuracy of time measurement. As a result of their heterogeneity, grain boundaries, cracks and pores, most of the rocks have such a high attenuation that only in a few cases it is possible to employ the extremely precise measuring techniques commonly used in solid state physics for single-crystal investigations, e.g. the phase comparison method of McSkimmin [25, 26]) and the CW techniques [27, 28]. A number of different electronic methods were developed for the time measurement

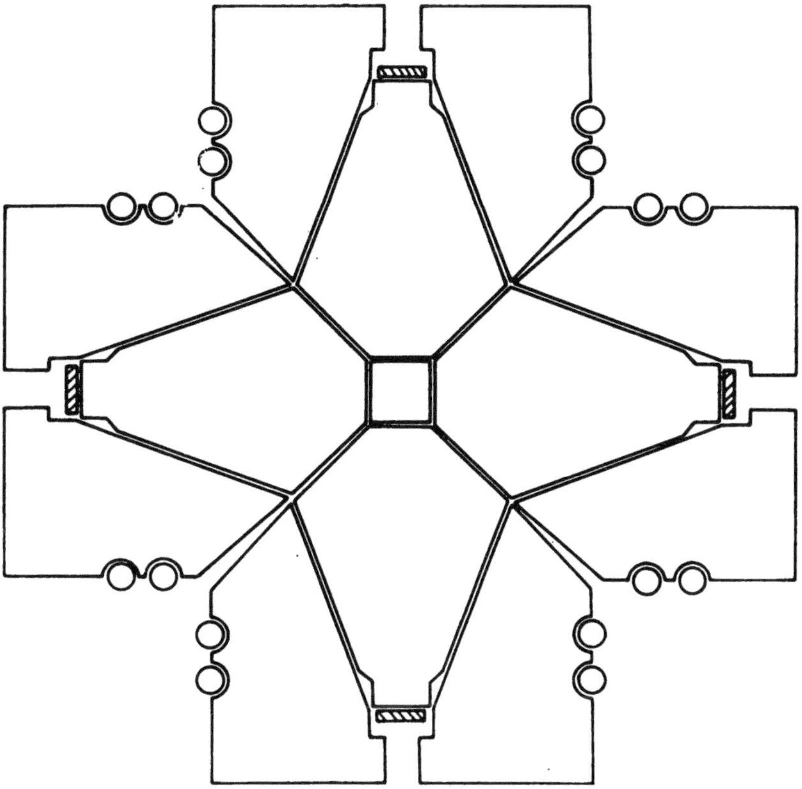

Fig. 1. Sectional view of four anvils and the sample.

in the pulse techniques which are in most cases used in transmission [29, 30, 31]. The benefits of a high accuracy and a convenient and rapid execution of the tests can be achieved by the use of electronic counters. With these counters a gate circuit is opened by the transmitting pulse and re-closed by the receiving pulse. The achievable accuracy depends on the upper limiting frequency of the counter used, because the desired delay time results from the number of pulses which passed the gate.

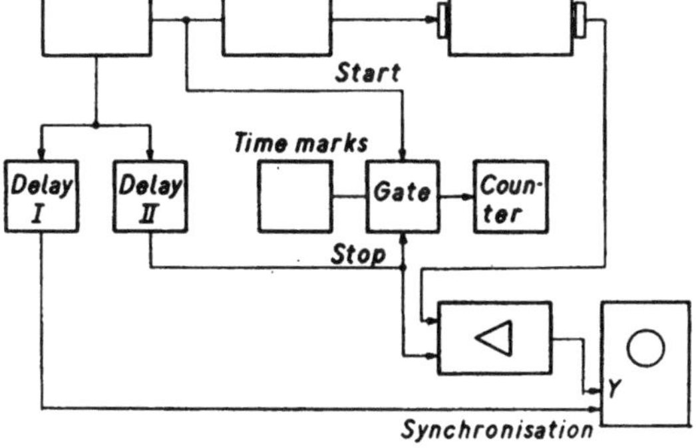

Fig. 2. Block diagram of the ultrasonic measuring device.

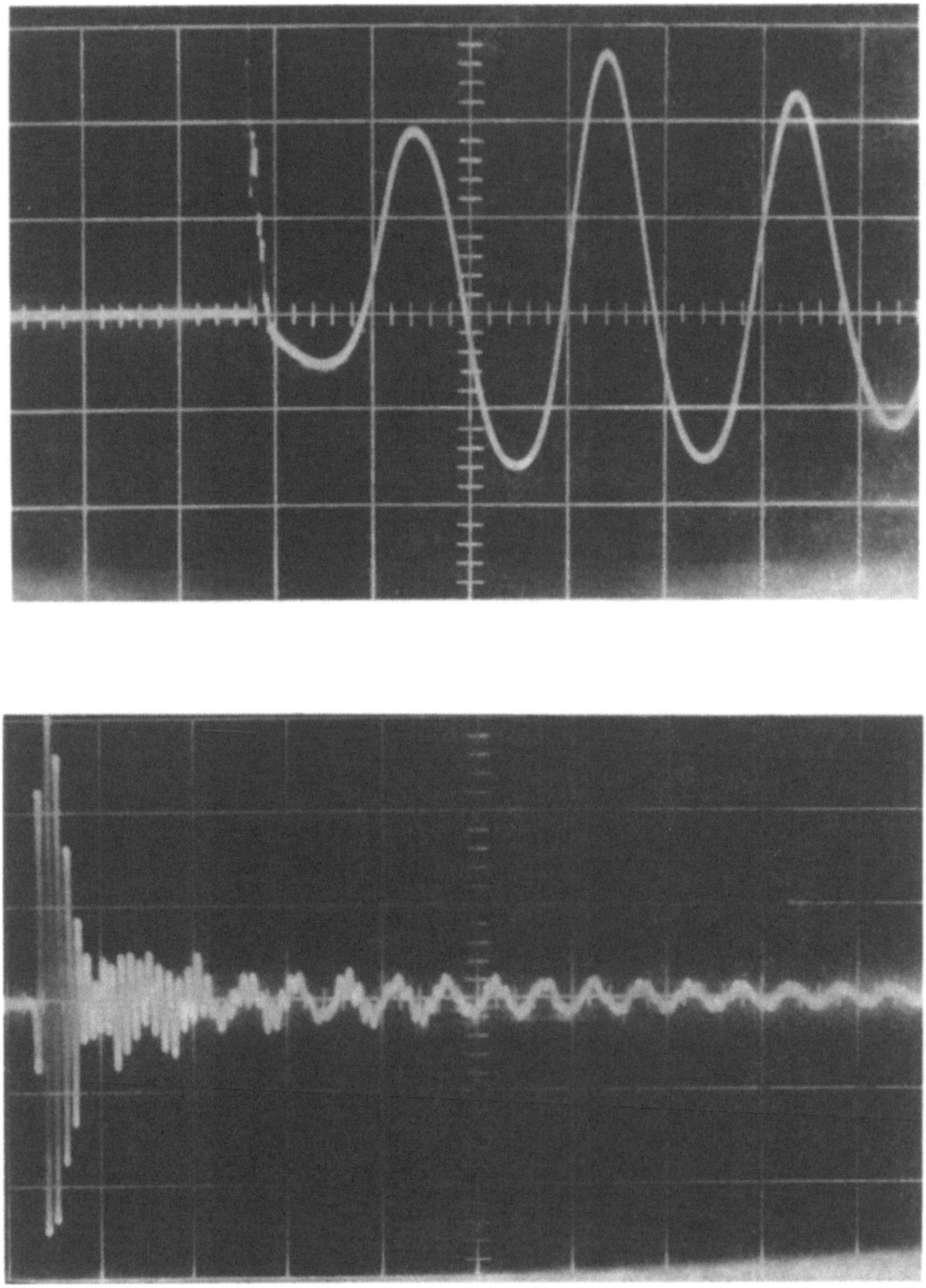

Fig. 3. Segments of the receiving signal with different time expansions
a) 0.5 μsec/cm with superimposed measuring mark; b) 10 μsec/cm.

For the measurements carried out according to this principle in the triaxial equipment, the Soviet counter frequency meter Tscha 3-34 was employed. It has an upper limiting frequency of 120 MHz, so that it can process pulses following one another in intervals of 10 nsec. In our opinion a much higher time resolution is not reasonable because the pulse onsets in rocks can rarely be localized with higher accuracy. The ultrasonic transducers were excited by electric pulses supplied by a pulse generator EMG-1157. The conversion into elastic pulses was effected by thickness vibrators made of Piezolan *S*, with a resonance frequency of 1 MHz. Since the receiving pulses were not always sufficient to inhibit the gate circuit, a special low-noise preamplifier was developed. The amplifier has a second input, by means of which a pulse mark can be mixed with the receiving signal. In practical operation a certain inadequacy of a fully-automatically operating gate circuit was encountered, which is also common to the sing-around technique [32]. The flip-flop of the gate circuit is inhibited at a well-defined, fixed level of the receiving signal. This signal value is completely independent of the shape of the first onset. If during the measurement, for example, by application of pressure to the sample, the signal shape or size vary, then the tripping time is shifted in relation to the leading edge of the first onset. A change in the gain of the input signal likewise leads to such a shift, and hence to a variation of the measured value.

Therefore it is useful to check the input signal during the measurement by means of an oscillograph and individually to determine the stopping of the counting operation. For this purpose, a measuring mark is mixed into the receiving signal by utilizing the second amplifier input. It is supplied by the pulse generator and can be delayed by an optional interval with respect to the transmitting pulse by means of the retarder I. This pulse, which simultaneously stops the counting operation, can thus be shifted to any point of the displayed receiving signal. Thus by choosing a suitable delay time it is possible to determine any desired point on the leading edge of the first onset, at which the gate circuit is inhibited, viz., to determine the end of the measuring time. This choice can be visually checked on the viewing screen. Fig. 2 shows the general wiring diagram of the measuring device.

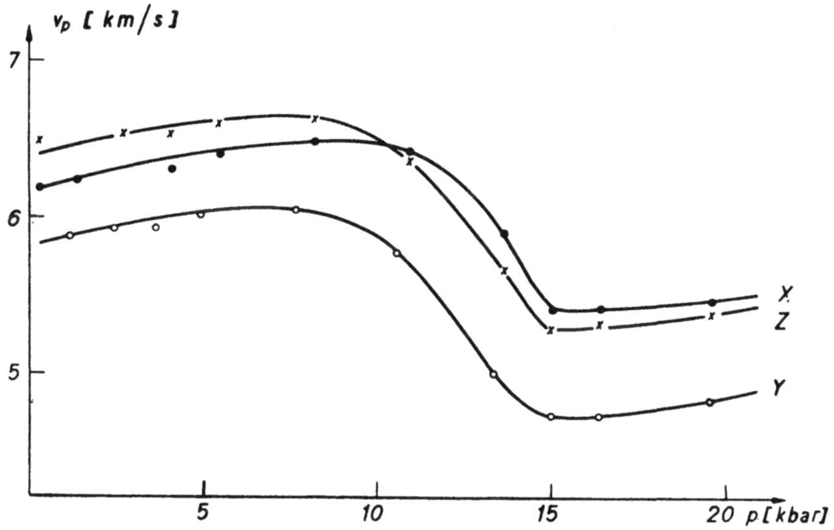

Fig. 4. Elastic anisotropy on the calcite/single-crystal phase transition.

A second retarder is used for delaying the sweep of the oscillograph. In this way it can be determined, which time interval (as referred to the transmitting pulse) shall be displayed on the screen. Thus it is possible to pick out any desired segment of the receiving signal. Depending on the selected time base setting, this interval can be displayed on the oscillograph with different degrees of expansion. Fig. 3 shows several segments of the same receiving signal with defferent time expansions.

The propagation velocities of compressional waves, v_p, can be calculated from the deformation and delay-time measurements after applying suitable corrections for the influence of the equipment and of the pyrophyllite on the results of measurement. By way of example, the calcite 1—2 transition was detected in a calcite single-crystal (see Fig. 4).

EFFECTS OF HIGH PRESSURE ON THE VELOCITY OF ELASTIC WAVES IN SALTS OF DIFFERENT CHEMICAL COMPOSITION

M. P. VOLAROVICH, E. I. BAYUK, and R. S. AKULOVA

Knowledge of the laws governing changes in the velocity of elastic waves and density of salt samples at high pressure makes it possible to interprete with more confidence acoustic logging data form the prospecting wells drilled in the Caspian depression and identitify different structures in the salt stratum. The salt sediments in this region are gerally represented by massive beds of rock salt (halite) or a mixture of halogenides with a different percent content of chemical components, or minor layers of salt interbedded with anhydrite rock and plastic clay. The halogenides most often include halite (NaCl), sylvine (KCl), carnallite ($KCl \cdot MgCl_2 \cdot 6H_2O$) and bischofite ($MgCl_2 \cdot 6H_2O$).

To interprete geophysical observations in regions of salt tectonics it is important to abtain information about the relationships between pressure and velocity of elastic waves in samples of these materials and of the rocks that are made up of them.

Except for work [33], the literature contains virtually no data on laboratory investigations of elastic properties of salt specimens under uniform pressure for such rocks as bischofite, carnallite and sylvine, though the elastic characteristics of rocks salt have been studied somewhat more thoroughly [33—37].

This paper deals with experimental studying the velocities of elastic waves at uniform pressure up to 10 kbars in salt samples of different mineral composition. Some of the sample were cut out of the cores of deep wells, some were made artificially from mixtures of polycrystalline powders with a certain chemical composition by compressing them at a pressure of 5 kbars by the technique described in [36]. In a high-pressure setup of the cylinder-piston type, the velocity of longitudinal waves was measured by a method of ultrasound pulse sounding and at the same time the density was determined by the piston displacement method [38]. Lead was used as a pressure-transmitting medium. The samples were 17.5 mm in diameter and 20 mm high.

Table 1 shows the percent content of principal minerals in the natural samples. Fig. 1 illustrates typical dependences of longitudinal wave velocities (v) for samples of halit, sylvine, carnallite and bischofite. As is seen from the dependence $v = f(p)$, the halite sample 22 h has the highest velocity of elastic waves and sylvine sample 6 s and carnallite sample 9 c have somewhat lower velocity values. The dependences $v = f(p)$ for these samples are arranged in the figure downwards in the order of decreasing density (Table 2). It was noted that when pressure rises from atmospheric to 1.5 kbar the velocity in all salts increases due to the closing of micro fissures and fissure pores, which renders the rock more uniform, increases the area of intergrain contacts in rock-forming

Table 1. Mineral Composition of Natural Samples

Sample No.	Well Name	Sampling depth, [meters]	Mineral Composition in weight percent					
			NaCl	KCl	$MgCl_2 \cdot 6H_2O$	$KCl \cdot 6H_2O$	$CaSO_4$	$CaSO_4 \cdot 2H_2O$
2 h*	No. 3 Svetoloyarskaya	1126—1130	99.0	—	—	—	0.29	—
4 h*	No. 20 Antipovsko-Balykleyskaya	1512—1517	85.91	—	1.45	12.26	0.29	—
21 h**	No. 101 Demidovskaya	2857—2863	100					
22 h*	No. 1084 Vikhlyan-tsevskaya	1350—1356	98.3	—	0.41	—	0.69	—
5 s*, 6 s*	No. 30 Antipovsko-Balykleyskaya	1507—1512	12.33	82.45	—	—	—	1.65
16 s**	No. 101 Demidovskaya	3297—3303	29	70	—	—	traces	—
9c*	No. 45 Antipovsko-Balykleyskaya	1385.8—1386	15.5	—	0.81	84.56	—	—
3b*	No. 45 Antipovsko-Balykleyskaya	1384.6—1384.8	—	—	97.17	1.23	1.13	—

* — analysis performed at Moscow University.
** — analysis performed at Volgograd Institute of Petroleum Prospecting.

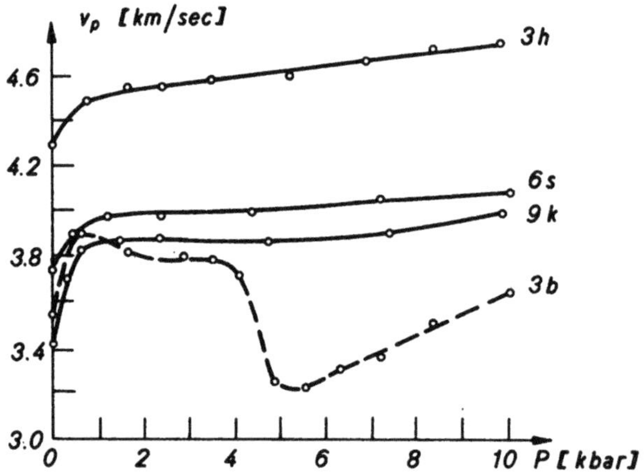

Fig. 1. Velocity of longitudinal waves versus pressure in salt samples. Plots numbered as in Table 1.

minerals and allows the energy of elastic waves to be transmitted more fully in salt samples. An increase in velocity within this pressure range for sample 22 h is 7% at the overall porosity coefficient $D = 0.41\%$, approximately 10% at $K = 2\%$ for sample 6 s and 19% at $K = 3.5\%$ for sample 9 c. The following regularity can thus be made out: the higher the overall porosity coefficient, the higher is the velocity increase within the pressure range form atmospheric to 1.5 kbar.

Table 2. Sample Density versus Pressure

Sample No.	Density, ϱ [g/cm^3]						
	Pressure [Kbar]						
	0.001	1	2	3	4	5	10
3 h	2.14	2.16	2.17	2.18	2.22	2.22	2.24
4 h	2.07	2.12	2.14	2.15	2.18	2.21	2.23
21 h	2.12	2.14	2.16	2.17	2.18	2.19	2.24
22 h	2.12	2.13	2.15	2.17	2.18	2.19	2.24
5 s and 6 s	1.94	1.96	1.99	2.01	2.03	2.05	2.11
16 s	2.07	2.09	2.10	2.11	2.11	—	—
9c	1.61	1.65	1.67	1.72	1.74	1.75	1.79
3b	1.54	1.57	1.59	1.60	1.62	1.64	1.89

Increase in loading up to a pressure of 1.5 kbar brings the rock back to the stressed state in which it was under the natural conditions.

With a further rise in pressure, the velocity changes negligibly and the rock shows virtually no micro fissures. The laboratory results of velocity measurements for pressures of 3 to 4 kbars agree well with the velocity values calculated from acoustic logging field data by the mean interval time law. This conclusion was drawn earlier when studying samples of rock salt from wells in the Pripyat depression [37]. A high short-time quasi-hydrostatic pressure of 3 to 4 kbars appears to compensate for the prolonged impact which the geostatic rock pressure corresponding to the depth of a rock in a bed exerts on salt rocks. At a pressure of above 5 kbars the velocity for some samples begins to increase somewhat higher than within the interval of 1.5 to 5 kbars, apparently because of the mechanism of compression of volumetric pores. In a study of sylvine samples subjected to higher pressures, a polymorphic transformation at 1.6 kbar was recorded by ultrasonic techniques [39].

The depencence $v = f(p)$ for bischofite (sample 3b) is peculiar and does not obey the laws obtained for other types of salt. It has an anomalous pattern of velocity change within the interval from 4 to 6 kbars, where the velocity drops abruptly by 20%, has a minimal value at 5.3 kbars, and then increases steadily. It was suggested in [33] that the anomalous velocity change is due to polymorphic transformation or to the release of water of crystallization which the molecules of bischofite begin to lose easily at 1.5 kbar, as evidenced by the experiment. This phenomenon has not been studied completely and is being investigated further.

Admixtures in the samples affect significantly the velocity of longitudinal waves. Fig. 2 shows the dependence $v = f(p)$ for the samples of halite of different mineral compositions (see Table 1). The dependences $v = f(p)$ for samples 3 h and 22 h occupy the upper part of the plot and as far as the velocity of elastic waves is concerned approach the dependence $v = f(p)$ for sample 3(*c*/*p*) prepared from chemically pure (*c*/*p*) fine crystalline powder NaCl. A certain difference in the velocity values for these samples can be explained by the fact that despite their almost indentical composition (see Table 1) the *c*/*p* sample has a densely compressed, fine crystalline structure, while samples 3 h and 22 h have a structure of real crystals having the 3-mm grain size.

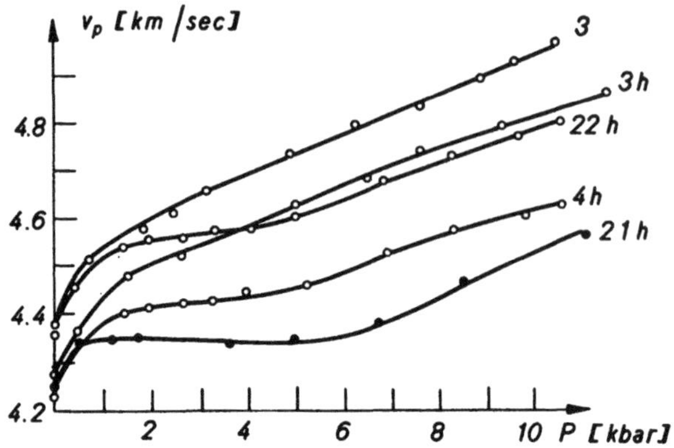

Fig. 2. Velocity of longitudinal waves versus pressure in halite samples.

The $v = f(p)$ curve for sample 4 h occupy the bottom part of Fig. 2 and has a lower velocity value. This sample contains only 86% halite, the remaining 24% being carnallite and bischofite which have a lower elastic wave velocity than halite. Sample 21 differs strikingly from the other halite samples. According to X-ray structure analysis data, it contains nearly 100% halite, but has a coarse crystalline structure, individual crystal grains being 10 mm or more in size, the larger halite grains as a rule containing pores in the form of blisters. For this reason the velocity value for sample 21 h is less than those for samples 3 h and 22 h.

Within the interval of 2 to 4 kbars the velocity of longitudinal waves in pure natural halite varies from 4.35 to 4.55 km/sec, which is in good agreement with the data of [34, 35].

Fig. 3 shows the $v = f(p)$ plots for sylvine samples. The curve for chemically pure sylvine does not occupy the upper position and lies below the curve for sample 16 s. The latter sample has a much greater velocity value than sample 10 (c/p) because, as analysis data show, it contains 30% halite admixture whose velocity and density are higher. The $v = f(p)$ plots for samples 10 (c/p) and 5 s are close to each other, even though sample 5 s contains 12.3% halite admixture which must increase the velocity in the sample. It will be interesting to compare these data with the velocity value in natural sylvine but, unfortunately, the latter is not available. Natural salts generally have a coarse-grain structure with volumetric pores and therefore the velocity in pure natural sylvine will be lower than in sample 10 (c/p).

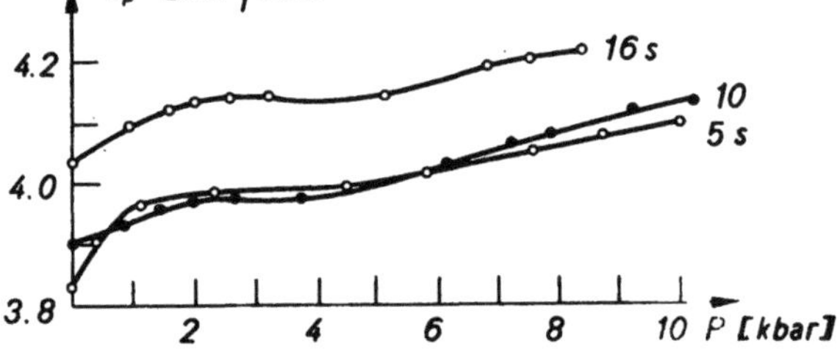

Fig. 3. Velocity of longitudinal waves versus pressure for sylvine samples.

It can be seen from Fig. 3 that halite admixtures tend to increase the velocity of longitudinal waves in sylvine.

Similar experiments were carried out on artificial samples of double mixture of Na, K, and Mg chlorides to ascertain the effects of the quantity and quality of admixtures in salts on the velocity of longitudinal waves in them. Samples of different mineral compositions were prepared from a mixture of "chemically pure" polycrystalline finely dispersed powders (Table 3) under indentical conditions which rule out the effect of grain size, fissure size, porosity, humidity, etc. on the velocity. The samples were pressed at a pressure of 5 kbar. The crystal grain size of the artificial samples was of the order of several micron fractions. The samples were arranged in three groups according to their mineral composition. The first group: KCl as the main rock-forming mineral and NaCl as the admixture mineral; the second group: $MgCl_2 \cdot 6H_2O$ as the main mineral and NaCl as the admixture mineral; the third group: $MgCl_2 \cdot 6H_2O$ as the main and KCl as the admixture mineral.

Fig 4. shows the plot of the velocity of longitudinal waves $v = f(p)$ and volume decrement $\frac{\Delta V}{V_0} = f(p)$ versus pressure for the first-group samples prepared from KCl and having different percentage contents of NaCl admixture. From the figure it is clear that the velocity of longitudinal waves is dependent on the percent content of admixture mineral and its density. The $v = f(p)$ plots for each sample are arranged in the order of increasing admixture content, i. e. downwards, and are bounded at the top by the $v = f(p)$ plot for pure NaCl and at the bottom by the same plot for pure KCl. The $\frac{\Delta V}{V_0} = f(p)$ plot conveys an idea about compressibility of the compound, i. e. about its structure. For this

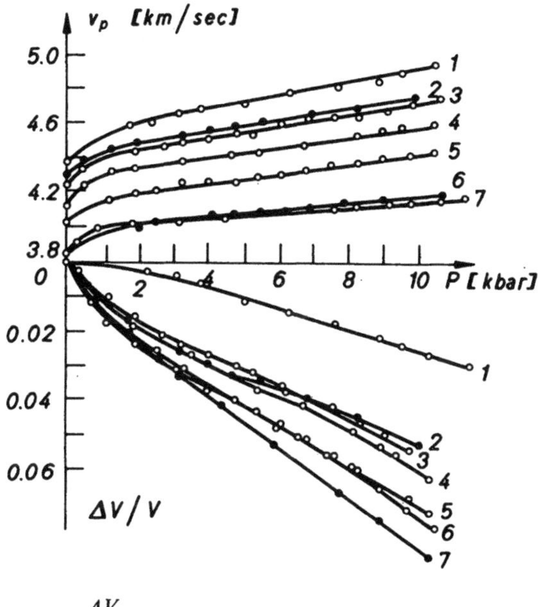

Fig. 4. $v = f(p)$ and $\frac{\Delta V}{V_0} = f(p)$ plots for different proportions of KCl and NaCl.
Plots numbered as in Table 3.

particular case the slope of the $\frac{\Delta V}{V_0} = f(p)$ for pure NaCl (sample 3, curve 1) is the lowest, after which the slope increases for each sample with an increase in the amount of KCl, i.e. admixtures of KCl makes the sample "less dense". Moreover, another feature is evident: the plots $\frac{v}{v_0} = f(p)$ for samples 106, 112 and 115 (curves 2, 3 and 4) containing 10%, 25% and 50% KCl respectively are arranged as one group, while the other plots with a higher slope form another group and tend toward the plot for pure KCl. It is possible to state that if KCl is regarded as the main rock-forming mineral and NaCl as the admixture, then the higher the percent content of NaCl, the higher are the velocity of elastic waves and the density and the lower is the slope of the volume decrement plot (Fig. 4, Table 1), i. e. NaCl admixtures make the structure denser and result in a considerable increase in the velocity of elastic waves.

For the second- and third-group samples, in which $MgCl_2 \cdot 6H_2O$ is the main mineral and NaCl and KCl are admixture, the plots $v = f(p)$ and $\frac{\Delta V}{V_0} = f(p)$ are shown in Figs 5 and 6 and the change of density versus pressure rise is presented in Table 3. The $v = f(p)$ plot

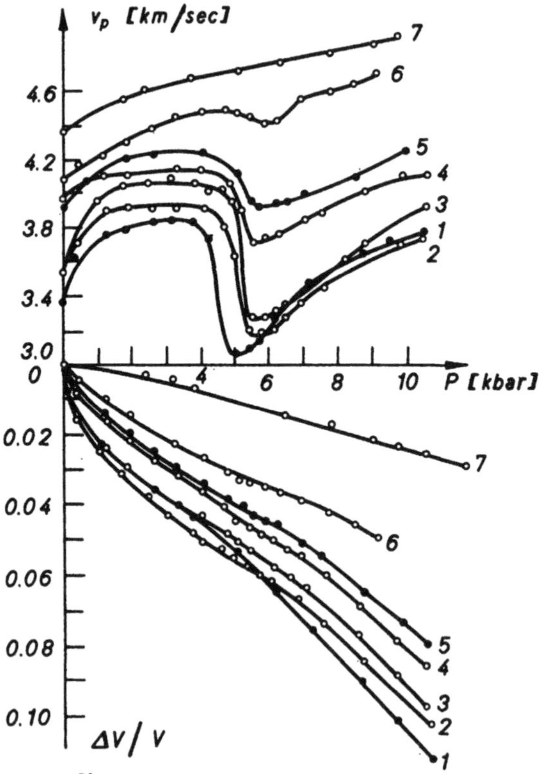

Fig. 5. $v = f(p)$ and $\frac{\Delta V}{V_0} = f(p)$ for different proportions of $MgCl_2 \cdot 6H_2O$ and NaCl. Plots numbered as in Table 3.

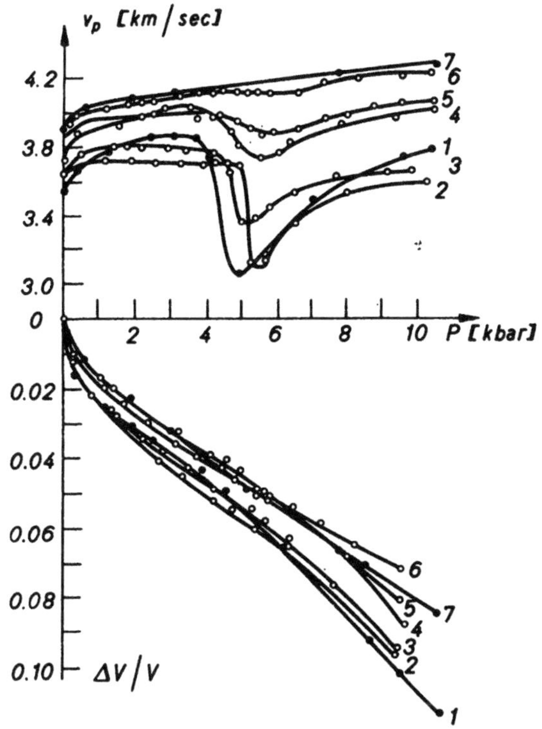

Fig. 6. $v = f(p)$ and $\frac{\Delta V}{V_0} = f(p)$ for different proportions of $MgCl_2 \cdot 6H_2O$ and KCl. Plots numbered as in Table 3.

shows an anomalous pattern within the range from 4 to 6 kbars due to special peculiarities of bischofite. NaCl and KCl admixtures gradually smooth out the anomalous range of values $v = f(p)$, depending on their percent content. NaCl admixtures increase velocity much higher than those of KCl and, in addition, the $v = f(p)$ plots for the second-group samples (Fig. 5) are arranged less densely than for the third-group samples (Fig. 5). This seems to be due to the fact that the difference in densities for NaCl (2.14 g/cm^3) and $MgCl_2 \cdot 6H_2O$ (1.58 g/cm^3) is higher than that for KCl (1.94 g/cm^3) and $MgCl_2 \cdot 6H_2O$. It should also be noted that an amount of KCl admixtures in excess of 50% smooths out the anomalous range of values $v = f(p)$ considerably, while a 10% KCl admixture almost does not change the usual pattern of the $v = f(p)$ plot for pure sylvine. For samples containing even a 10% NaCl admixture the anomalous range for the $v = f(p)$ plot remains, though much simplified.

Moreover, NaCl admixtures in the main rock-forming mineral, bischofite, gradually shift the flex point of the $v = f(p)$ plot or the minimum velocity plot into the higher pressure area. An increase in the percent content of NaCl admixtures leads to an increase in the velocity of elastic waves and the $r = f(p)$ plots are grouped in Fig. 5 downwards correspondingly to the amount of admixtures. The minimum velocity values shift from left to right.

The same is true of the third-group samples in which $MgCl_2 \cdot 6H_2O$ is the main rock-forming mineral and KCl the admixture, though for this group the velocity does not

Table 3. Mineral Composition of Artificial Samples of Chlorides of Na, K and Mg
Density of Samples at Pressures up to 10 kbars

Plot No. in Fig.	KCl as main mineral						$MgCl_2 \cdot 6H_2O$ as main mineral						$MgCl_2 \cdot 6H_2O$ as main mineral					
	NaCl admixture %	Density, g/cm³					NaCl admixture %	Density, g/cm³					KCl admixture %	Density, g/cm³				
		$P=0.001$	$P=1.5$	$P=3$	$P=5$	$P=10$		$P=0.001$	$P=1.5$	$P=3$	$P=5$	$P=10$		$P=0.001$	$P=1.5$	$P=3$	$P=5$	$P=10$
1	100, Smpl No. 3	2.14	2.16	2.17	2.19	2.25	0, Smpl No. 11	1.58	1.605	1.63	1.66	1.74	0, Smpl No. 11	1.58	1.605	1.63	1.66	1.74
2	90, Smpl No. 106	2.12	2.145	2.15	2.18	2.235	10, Smpl No. 97	1.59	1.62	1.65	1.63	1.76	10, Smpl No. 79	1.57	1.59	1.605	1.63	1.71
3	70, Smpl No. 112	2.07	2.095	2.115	2.14	2.20	25, Smpl No. 94	1.665	1.685	1.71	1.735	1.81	25, Smpl No. 74	1.62	1.66	1.68	1.705	1.81
4	50, Smpl No. 115	2.05	2.08	2.10	2.12	2.175	50, Smpl No. 91	1.72	1.76	1.78	1.80	1.875	50, Smpl No. 71	1.73	1.76	1.79	1.825	1.915
5	25, Smpl No. 119	2.0	2.045	2.065	2.09	2.155	70, Smpl No. 87	1.87	1.92	1.94	1.97	2.04	70, Smpl No. 67	1.81	1.84	1.86	1.89	1.97
6	10, Smpl No. 26	1.97	2.01	2.03	2.065	2.14	90, Smpl No. 83	2.04	2.06	2.08	2.105	2.16	90, Smpl No. 63	1.90	1.935	1.95	1.99	2.05
7	0, Smpl No. 10	1.94	1.96	1.985	2.005	2.10	100, Smpl No. 3	2.14	2.16	2.17	2.19	2.25	100, Smpl No. 10	1.94	1.973	1.988	2.012	2.102

increase with a rise in the percent content of the admixture mineral as much as for the preceding group. This may be due to the lower ratio between the densities of the two minerals.

The volume decrement plot $\frac{\Delta V}{V_0} = f(p)$ for the samples of the two groups also shows a peculiarity of its own. Thus, for $MgCl_2 \cdot 6H_2O$ samples with NaCl admixture the $\frac{\Delta V}{V_0} = f(p)$ plots become arranged in such an order that sample 3 of pure NaCl has the lowest slope and then the slope of $\frac{\Delta V}{V_0} = f(p)$ increases with a decrease in the percent content of NaCl admixtures. For $MgCl_2 \cdot 6H_2O$ samples with KCl admixture the $\frac{\Delta V}{V_0} = f(p)$ plots are arranged as in the previous case in the order of increasing percent content of KCl, i.e. as the amount of the admixture increases the $\frac{\Delta V}{V_0} = f(p)$ plots become arranged upwards in Fig. 6 and more densely than in the preceding case. It is remarkable that for the second-group and third-group samples the $\frac{\Delta V}{V_0} = f(p)$ plots have a different slope in the pressure range of 5 to 6 kbars, i. e. in the pressure range characterized by an anomalous change in the velocity of elastic waves, Consequently, KCl admixtures with a lower density ($p = 1.98$ g/cm^3) affect the structure of the substance less than NaCl admixtures with a higher density ($p = 2.14$ g/cm^3).

ELASTIC PROPERTIES OF SOME ROCK-FORMING MINERALS AND MINERAL MATTER DEFORMATION AT HIGH HYDROSTATIC PRESSURE

T. S. Lebedev, V. A. Korchin, Yu. P. Orovetsky, P. A. Burtny

Studies of physical properties of rock-forming minerals under high thermodynamic conditions contribute greatly to our knowldge of the character and values of changes in physical parameters of rocks in the Earth's interior. Physical properties of rocks are defined mainly by the properties of their constituent minerals. The knowledge of the elastic characteristics of minerals, their change with pressure and temperature and mineral composition of rock enables to calculate the elastic characteristics of multicomponent mineral associations in any thermodynamic conditions [41, 42 et al.].

It is possible thus to make a more detailed examination of the effect of pores, fissures, grain boundaries and other disturbances in structural homogeneity on a change of elastic parameters of rocks with depth [40—42, 57 et al.].

Minerals are formed under widely varying physical, chemical and thermodynamic conditions of deep and subsurface zones of the Earth. Certain minerals and associations bear the records of the regional environments in which they have been formed [58 et al.]. Therefore it is difficult to compare the results of the elastic parameters measurements in natural minerals sampled from different geological provinces and these results cannot always be used in discussing the state of rocks in a specific geologic region [40, 42, 53, 55, 57, 58 et al.].

We have studied elastic properties of certain rock-forming minerals sampled in different areas of the Ukrainian Shield. Those are quartz, potash feldspar, calcite, plagioclase, biotite, albitite, quartzite. Quartz (the Volynsky morion) is twinned with abundance of gas-fluid inclusions. Potash feldspar is represented by microcline-pertite and microcline. Quartz forms rather large crystals, potash feldspar is sampled from pegmatite veins in granitoids. Plagioclase is represented by polysynthetically twinned andesine (No 46) and labradorite (No 56). Biotite has the form of a large tabular crystal and is sampled from schlieren impregnated in granitoids. Quartzite is a single-mineral rock composed almost completely of quartz. All minerals, except for quartz, were sampled from genetically related rocks. This hampers their pre-experimental treatment and is the reason of the presence of various (though small) admixtures. The major constituent minerals of granites (quartz, plagioclase, microcline and biotite) were studied at high hydrostatic pressure to 14—15 kbar in different directions. Elastic velocities were studied with regard to the elements of crystal symmetry. The displacement components either coincided with the wave path (a pure longitudinal wave V_p) or crossed this direction (a pure shear wave V_s).

The results of the studies of $V_{p,s} = f(P)$ in different directions in quartz, plagioclase, microcline and biotite are presented in fig. 1.

The highest V_p values of 8.24 km/sec at atmospheric pressure were found in andesine No 46 along the line [010], the lowest — in biotite along [001] (3.35 km/sec). In the same condition, the lowest V_s value (1.85 km/sec) is also characteristic of biotite (1.85 km/sec) whereas the highest V_s (4.35 km/sec) was observed in quartz along the sixfold axis.

With increasing pressure all the minerals show a common regularity. It is a rapid increase of velocity in the initial range of pressure. Further increase of pressure causes a linear change of $V_{p,s} = f(P)$. Most scientists [40, 42, 57, 58 et al.] explain the sharp increase of elastic parameters at lower pressure by a change or smoothing of structural

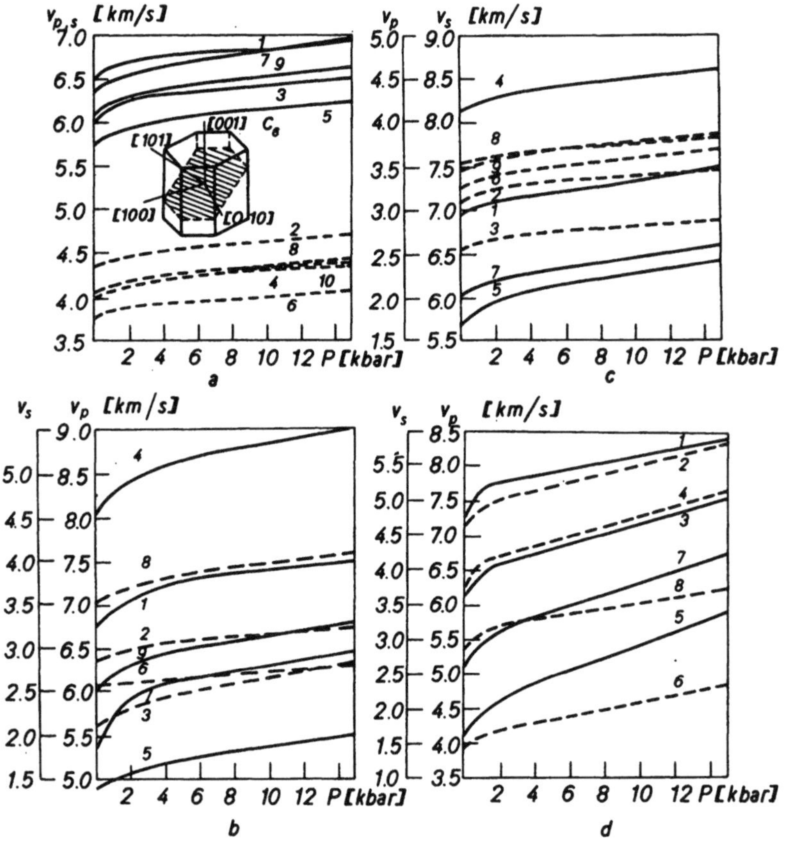

Fig. 1. Changes of V_p, $V_s = f(P)$ in some rock-forming minerals in various crystallographic directions. a — quartz: 1 — V_p [001], 2 — V_s [001] [010], 3 — V_p [010], 4 — V_s [010] [100], 5 — V_p [100], 6 — V_s [100] [001], 7 — V_p [101], 8 — V_s [101] [010], 9 — $\bar{V}_p$, 10 — $\bar{V}_s$; b — microcline: 1 — V_p [001], 2 — V_s [001] [100], 3 — [001] [010], 4 — V_p [010], 5 — V_p [101], 6 — V_s [101] [010], 7 — V_p [100], 8 — V_s [100] [010], 9 — V_s [100] [001], c — plagioclase: 1 — V_p [001], 2 — V_s [001] [100], 3 — V_s [001] [010], 4 — V_p [010], 5 — V_p [101], 6 — V_s [101] [010], 7 — V_p [100], 8 — V_s [100] [010], 9 — V_s [100] [001]; d — biotite: 1 — V_p [010], 2 — V_s [010] [100], 3 — V_p [101], 4 — V_s [101] [010], 5 — V_p [001], 6 — V_s [001] [100], 7 — $\bar{V}_p$, 8 — $\bar{V}_s$. The first values bracketed indicate the direction of the wave which is propagating, the second show its displacements. $\bar{V}_p$, $\bar{V}_s$ are the velocities calculated on the basis of the averaged values of the elastic modules.

defects in minerals, i.e. by microfractures, ultramicropores, foreign phase inclusions, local clusters of admixed atoms and various defects in crystal lattice (twins, dislocations, point defects).

Without consideration of the original defective state of minerals to 2—3 kbar, let us examine only a linear change of velocities $V_{p,s} = f(P)$. The most significant change in mean $V_{p,s}$ values is characteristic of biotite ($dV_p/dP = 0.083$ km/s kbar; $dV_s/dP = 0.048$ km/s kbar), the smallest change — of plagioclase (andesine No 46) and quartz ($dV_p/dP = 0.012$; 0.025 km/sec kbar and $dV_s/dP = 0.01$; 0.012 km/sec kbar, respectively).

It has been found that of all the minerals studied the highest anisotropy is characteristic of biotite. In biotite V_p anisotropy decreases with pressure (dV_p/dP = is higher along cleavage than across), the V_s anisotropy behaves reversely $\left(\frac{dV_s \perp \text{cl.}}{dP} < \frac{dV_s \parallel \text{cl.}}{dP}\right)$ (fig. 1, d).

Quartz shows a lower degree of anisotropy. With increasing pressure the anisotropy along crystallographic lines remains almost unchanged (fig. 1, a). The same behaviour is observed in V_p and V_s anisotropy of plagioclase and microcline (fig. 1, b, c), but only at $P > 2$ kbar. Below 2 kbar the velocity anisotropy changes rapidly, especially in microcline (fig. 1, b). This phenomenon is evidently caused by an anisotropic measurement of the mineral structure defects whose effect on acoustical properties decreases with pressure.

With growing pressure, changes in elasticity of crystals is less affected by the defects than by a relative positions of atoms within the crystalline lattice. The high pressure effect on

Table 1. Moduli of elasticity (C_{ij}) and constants of pliability (S_{ij}) in Quartz (a) and Biotite (b) at different hydrostatic Pressures

a						b					
P, kbar	11	12	13	33	44	P, kbar	11	12	13	33	44
0.001	0.878*	0.128	0.132	1.126	0.502	0.001	1.752*	0.354	0.127	0.556	0.068
	1.179**	−0.153	−0.120	0.916	1.992		0.602**	−0.113	−0.112	1.849	14.706
0.5	0.906	0.134	0.133	1.168	0.516	0.5	1.859	0.367	0.132	0.611	0.077
	1.143	−0.152	−0.113	0.882	1.938		0.566	−0.105	−0.100	1.679	12.987
1.0	0.921	0.137	0.135	1.186	0.523	1.0	1.935	0.379	0.130	0.647	0.085
	1.125	−0.151	−0.111	0.868	1.912		0.542	−0.100	−0.089	1.581	11.764
2.0	0.943	0.145	0.141	1.200	0.535	2.0	1.986	0.332	0.190	0.699	0.093
	1.101	−0.152	−0.111	0.859	1.869		0.528	−0.077	−0.122	1.497	10.753
3.0	0.959	0.151	0.147	1.207	0.542	3.0	2.011	0.309	0.239	0.746	0.096
	1.084	−0.153	−0.113	0.856	1.845		0.524	−0.063	−0.148	1.435	10.417
5.0	0.983	0.155	0.159	1.229	0.554	5.0	2.063	0.293	0.294	0.802	0.110
	1.060	−0.148	−0.112	0.844	1.805		0.516	−0.049	−0.171	1.372	9.091
7.5	0.997	0.155	0.174	1.245	0.563	7.5	2.129	0.237	0.347	0.877	0.126
	1.048	−0.141	−0.126	0.840	1.776		0.503	−0.025	−0.189	1.290	7.937
10.0	1.011	0.151	0.189	1.261	0.574	10.0	2.196	0.196	0.399	0.954	0.142
	1.034	−0.129	−0.136	0.834	1.742		0.492	−0.007	−0.201	1.204	7.042
15.0	1.040	0.144	0.219	1.296	0.596	15.0	2.333	0.109	0.499	1.131	0.179
	1.008	−0.108	−0.152	0.823	1.678		0.474	−0.025	−0.220	1.079	5.587

* C_i value in Mbar
** S_{ij} value in Mbar^{-1}

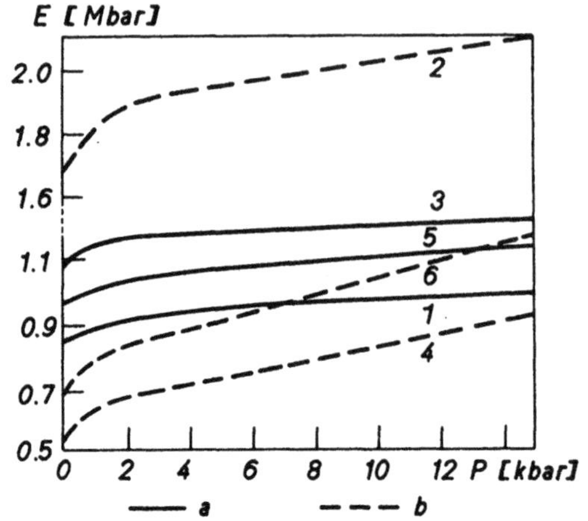

Fig. 2. Changes of $E = f(P)$ for a single crystal of quartz (a) and biotite (b). 1, 2 — $E_{11} = \frac{1}{S_{11}}$ along [100], 3, 4 — $E_{33} = \frac{1}{S_{33}}$ along [001], assuming that the tin core is deformed and all deformations occur without hindrance; 5, 6 — are calculated by averaging formulae for hexagonal crystals.

the change of elastic parameters in connection with the lattice elasticity varies not only with minerals but with crystallographic lines within the mineral (fig. 1). The latter produces a complex stress structure within the crystal volumes. So, let us consider more details about anisotropic changes in elastic characteristics of crystals. The phenomena are reflected in variations of independent values of pliability constant (S_{ij}) and rigidity modulus (C_{ij}) of the mineral. In quartz and biotite, (by its elastic properties, the latter may by considered a hexagonal crystal with [001] line orthogonal to the cleavage plane), the velocities

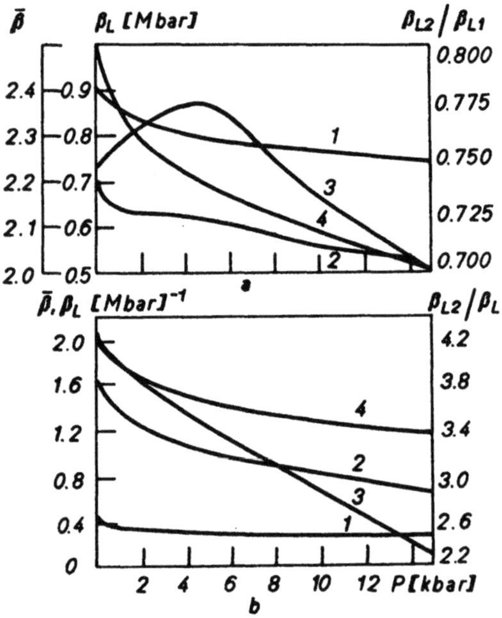

Fig. 3. Changes of $\beta = f(P)$ for single crystals of quartz (a) and biotite (b). 1 — linear compressibility (β_{L1}) along the axis 001, 2 — β_{L2} in the ⊥ 001 plane, 3 — $\beta_{L2}/\beta_{L1} = f(P)$, 4 — volume compressibility calculated on the basis of the averaged modules.

required for calculation of elastic parameters were measured. The values of C_{ij} and S_{ij} calculated for different hydrostatic pressure (to 15 kbar) are shown in Table 1. In microcline and plagioclase velocities were measured only in some directions (fig. 1, b, c). So, for these minerals only nine elastic parameters could be calculated.

The data on the values of linear compressibility, Young modulus, Poisson coefficient calculated for different crystallographic lines and presented, in terms of elastic parameters, as $\beta_\alpha = (S_{11} + S_{12} + S_{13}) - (S_{11} + S_{12} - S_{13} - S_{33}) \cdot \gamma^2$ (γ — direction cosine for the axis of symmetry), $E_i = 1/S_{ij}$ ($i = 1, 2, 3$), $\sigma = S_{ij}/S_{ii}$; as well as elastic characteristics calculated for polymineral aggregates by the averaging formulas of Vogt-Reuss-Hill [41], allow to judge on complex processes (elastic, brittle, plastic, etc.) operating in the mineral matter under high hydrostatic pressures.

The changes of Young modulus, compressibility and Poisson's ratio with pressure for quartz and biotite are presented in fig. 2—4 and Table 2.

Elastic properties of minerals in natural conditions are dependent, besides the atomic and molecular interaction, on internal remanent stresses closely related to formation and evolution of the mineral. Distinguished are macro- and microstresses characterizing single minerals and polycrystalline associations. In physics of solids remanent stresses are considered to be caused by disarrangements in the crystal lattice. Their values and signs depend mainly on density of homogeneous dislocations. Microstresses are local remanent stresses observed in microzones. They originate in polycrystalline associations during formation of mineral associations from interactions between grains. Remanent microstresses are caused mainly by phase transformations, temperature variations, anisotropy of mechanical properties in single grains, grain boundaries and disintegration of grains into fragments and blocks resulted from deformation. In other words, remanent microstresses result from the significant inhomogeneity and anisotropic properties of the crystal material responsible for the specific defect structure. Remanent stresses have a mechanical effect on the rock

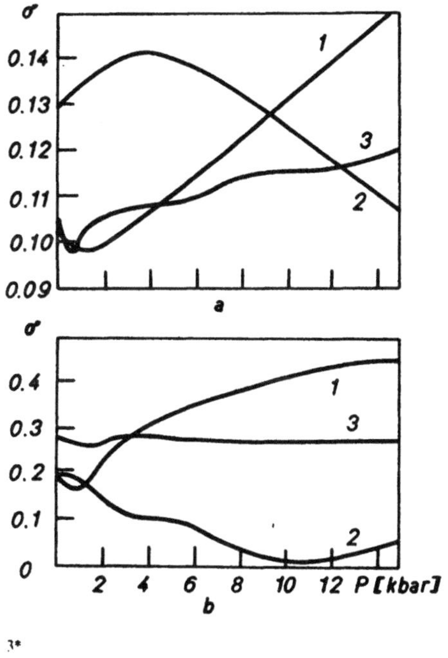

Fig. 4. Changes of $\sigma = f(P)$ for single crystals of quartz (a) and biotite (b).
Deformation ratio along the [100] and [001] axes (S_{13}/S_{11}) — 1, 2 — deformation ratio along [100] and [010] (S_{12}/S_{11}) assuming that the uniaxial stres acts along [100], 3 — mean value, calculated on the basis of the averaged modules.

Table 2. Average values of elastic Characteristics of Quartz (a) and Biotite (b) at different Pressures

	P, kbar	Averaging after Voigt			Averaging after Reuss			Average values (averaging after VRH)						
		G_V	K_V	E_V	G_R	K_R	E_R	$\bar{G}$	$\bar{K}$	$\bar{E}$	$\bar{\sigma}$	$\bar{\beta}$	$\bar{V}_p$	$\bar{V}_s$
		Mbar										Mbar	km/sec	
a	0.001	0.442	0.407	0.974	0.433	0.401	0.955	0.437	0.404	0.964	0.103	2.48	6.08	4.05
	0.5	0.456	0.406	0.995	0.446	0.414	0.984	0.451	0.410	0.990	0.098	2.44	6.16	4.11
	1.0	0.462	0.427	1.019	0.453	0.421	1.000	0.457	0.424	1.009	0.104	2.36	6.22	4.14
	2.0	0.471	0.438	1.040	0.461	0.432	1.020	0.466	0.435	1.030	0.105	2.30	6.29	4.18
	3.0	0.476	0.446	1.053	0.467	0.441	1.035	0.471	0.443	1.043	0.107	2.26	6.33	4.20
	5.0	0.486	0.460	1.078	0.477	0.450	1.057	0.481	0.455	1.067	0.109	2.20	6.41	4.25
	7.5	0.492	0.472	1.095	0.482	0.465	1.075	0.487	0.468	1.085	0.114	2.14	6.47	4.27
	10.0	0.499	0.482	1.113	0.489	0.475	1.092	0.494	0.478	1.102	0.116	2.09	6.53	4.31
	15.0	0.514	0.504	1.151	0.503	0.496	1.128	0.508	0.500	1.138	0.121	2.00	6.64	4.37
b	0.001	0.397	0.586	0.972	0.141	0.420	0.380	0.269	0.503	0.683	0.270	1.99	5.09	2.85
	0.5	0.427	0.621	1.042	0.159	0.454	0.427	0.293	0.538	0.744	0.270	1.86	5.30	2.98
	1.0	0.448	0.644	1.091	0.174	0.474	0.465	0.311	0.559	0.787	0.265	1.79	5.43	3.07
	2.0	0.467	0.677	1.139	0.188	0.523	0.504	0.328	0.600	0.832	0.268	1.67	5.60	3.15
	3.0	0.474	0.705	1.162	0.194	0.566	0.522	0.334	0.636	0.853	0.276	1.57	5.72	3.18
	5.0	0.491	0.743	1.207	0.217	0.617	0.583	0.354	0.680	0.905	0.278	1.47	5.90	3.27
	7.5	0.520	0.777	1.275	0.244	0.671	0.653	0.382	0.724	0.975	0.276	1.38	6.11	3.40
	10.0	0.548	0.816	1.343	0.270	0.730	0.721	0.409	0.773	1.043	0.275	1.29	6.32	3.52
	15.0	0.607	0.890	1.484	0.326	0.835	0.865	0.467	0.863	1.187	0.271	1.16	6.70	3.76

similar to the effect of useful loading. When the solid is affected from outside, the remanent stresses interact with the stresses from external loading. This brings about a change in the causes of the remanent stresses.

The contribution of the inhomogeneity to the stress distribution within the mineral and a change of the defect structure at hydrostatic pressure have been thoroughly studied in a crystal of quartz [48]. The quartz (morion) crystals formed a heterogeneous system characterized by numerous gasfluid inclusions and the presence of the Brazilian twins. The experiments were performed on prismatic samples of $14 \times 12 \times 4$ mm. Those were separately inserted into the chamber and extracted therefrom at discrete pressures of 0,5; 1; 2; 3; 5; 8; 12 and 15 kbar.

Before experiment a microscope would reveal optical inhomogeneties responsible for a specific pattern interfering with transparency of the crystal. After experiment, the optical inhomogeneities disappeared. Remanent deformation of the crystal was established by measuring linear dimensions of the crystal at the pressure values listed above. The peak value of remanent deformation was found in a range of 0.001—5 kbar. No linear deformation was revealed with further increase of pressure to 15 kbar (fig. 4). The refraction index of quartz remained the same before and after experiment. Comparison between lauegrams at 0.001, 5 and 15 kbar revealed no asterism. The maximum turn of direction angles of diffraction was observed at 5 kbar. The samples which had undergone the pressure effect showed through microscope and induced jointing characterized by development of linear microcataclastic zones and echelon fractures striking at different angles about the axis of these zones [48].

In quartz under hydrostatic pressure, we have thus identified apart from the elastic two kinds of deformation [50, 59]. Plastic deformation is characterized by remanent changes (fig. 4). Brittle deformation is revealed in microcataclastic zones fringed by fissures [48].

As known, the mechanism of the solid deformation is driven by redistribution of the dislocation density under the effect of internal stresses [40, 43, 45, 50, 51, 52, 54 et al.]. For this reason we have studied the dislocation structure in morion samples before and after application of pressure. To identify the dislocations the method of etching the quartz crystal in hydrofluoric acid is adopted [53, 56]. In the samples which have not been affected by pressure numerous single linear and radial exits of dislocations are found (fig. 5). Beyond 3 kbar the amount of etching patterns decreases and at 5 kbar and further they disappear completely (fig. 6) [48].

It is established experimentally [48] that, for quartz there exist a threshold pressure of 3—5 kbar which causes irreversible changes in the crystal such as removal of optical

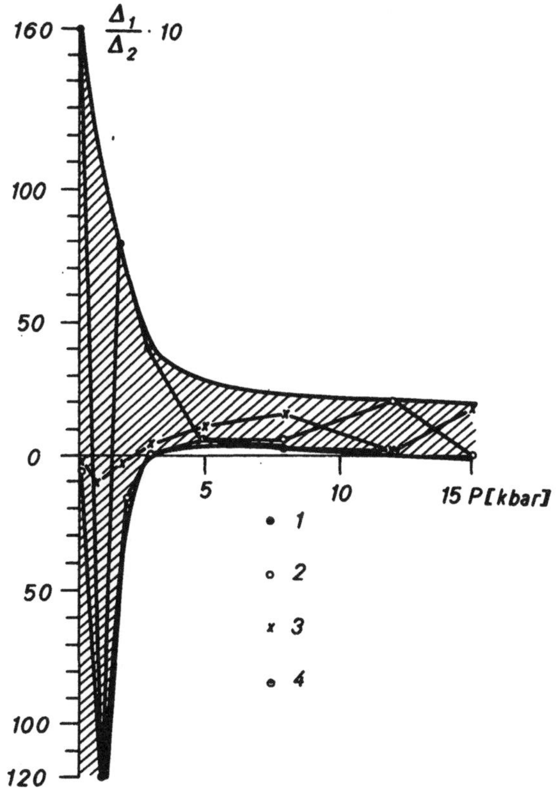

Fig. 5. Changes of linear dimensions of the quartz crystal at pressure.
1 — long axis to short axis ratio, 2 — long axis to transverse ratio, 3 — short axis to transverse ratio, 4 — common points.

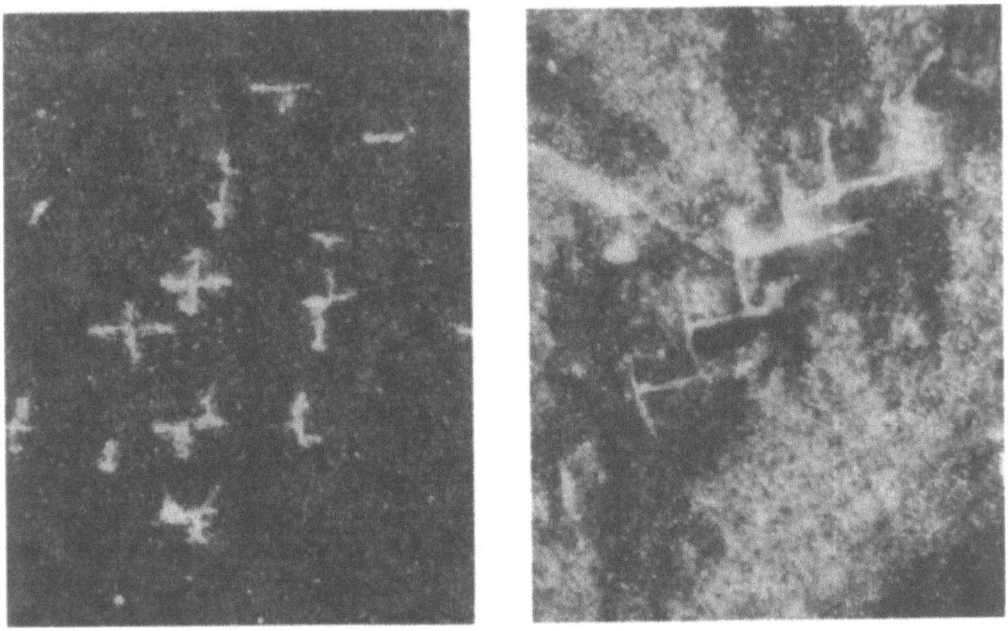

Fig. 6. Patterns of pyramidal etch cavities in quartz at the points of dislocation exits. Thin section, etched surface, oblique reflected light.
a — 250-fold magnification; b — 575-fold magnification.

inhomogeneities and disappearance of the pyramidal etching patterns directly related to exits of dislocations. These leads to perfection of the crystal structure of morion. Hydrostatic pressure thus originates in the crystal inhomogeneous stresses resulting in displacements of dislocations, possibly, by way of translation. The dislocations tend to penetrate into the areas of the highest stresses [43, 44, 51, 52, 54, 56] such as the crystal surface, fissures and local clusters of inhomogeneities. Or, the dislocations are annihilated in the common atomic plane. In the inhomogeneous crystal both mechanisms may work simultaneously. This is justified by a constancy of refraction index and the absence of asterism along with the existence of remanent deformation zones [43, 44, 52].

Simultaneous studies of elastic characteristics along different crystallographic lines of the crystal by the ultrasonic method confirm our assumption that, under hydrostatic compression, an anisotropic accumulation of elastic energy occurs.

The velocities measured in different directions were used in calculating moduli of elasticity characterizing the coupling forces and binding energy in the solid. Fig. 3—4 show the results of the Poisson's ratio and compressibility calculations for orthogonal lines in quartz. There is an increase of σ on certain planes of the crystal. This, by itself, causes the activation of dislocations mobility in glide planes. Since an increase of Poisson's ratio in the crystal as stated previously [45, 54] widens the marginal dislocation, compression within the atomic extraplane will push the dislocation out of its glide plane. In other words, the increase of Poisson's ratio characterize a growth of plasticity of the crystal thus favouring translations.

Comparing σ and β values obtained along different lines in morion (fig. 3—4) proves that, at hydrostatic pressure, there is a clear anisotropy in the change of elastic parameters along crystallographic lines, which brings about a nonuniform distribution of internal stresses causing disturbance in equilibrium state of the dislocation and, hence, the change in the defect structure of the crystal [57]. So, with pressure growing to 5—6 kbar, the σ-value along one of lines first increases and then, beyond 6 kbar, decreases, thus characterizing a decrease in plasticity of the system or increase of its brittleness. At the same time, the σ values measured in the plane perpendicular to the line show that after their decrease to 2.5 kbar they start increasing. The different behavior of $\sigma = f(P)$ (as well as that of β) along different lines characterize the internal inhomogeneity of stresses within the crystal volume. As seen on the graphs, an anomalous behaviour of σ and β can be detected at pressures from the atmospheric to 2.5 kbar and reach a maximum at 3—5 kbar. The hydrostatic pressure of 3—5 kbar is evidently most favourable for activation of the displacement of dislocations. The further pressure increase does not yield essential change in anisotropy of elastic stresses of the crystal. The decrease of σ beyond 6 kbar along certain lines characterizes a tendency of the crystal to adopt a more brittle state [47]. This, in turn, favours a development of the elements of brittle destruction within the crystal. These are observed experimentally in sections of samples under maximum pressures.

Summarizing all the above facts we may state that, at high hydrostatic pressure, the single crystal of quartz with natural defect structure acquires a complex stressed state. Under sufficient dynamic loading the defects (dislocations in particular) are redistributed within the crystal volume. To 3—5 kbar, the quartz crystal is characterized by the highest anisotropy of its elastic properties but upon reaching this threshold pressure the morion

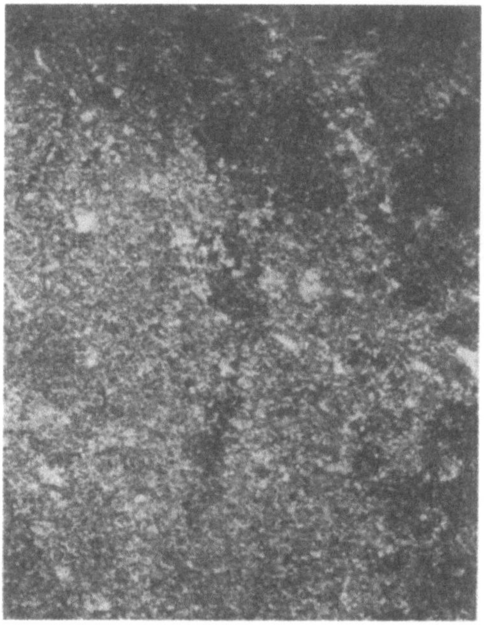

Fig. 7. Quartz free of dislocations at 5 kbar. Thin section, etched surface, oblique reflected light, 225-fold magnification.

crystal in local (less defective areas) gets free of the dislocations. With further increase of pressure brittleness of the mineral grows, the balanced state of large localizations of defects is being disturbed and a possibility arises for brittle deformations of the mineral recorded as newly formed microfissures.

Similar results have been obtained in artifical single crystal of garnet ($Y_3Fe_{5-x}Ga_xO_{12}$ and $Y_3Fe_5O_{12}$) [46]. The method of ferromagnetic resonance was applied to studying the superfine defect structure of single crystals after the effect of high quasihydrostatic pressure and temperature. It has turned out that there are critical areas of pressure, temperature, dislocation density and amount of admixed atoms in the crystal where a quantitative redistribution (decrease or increase) of defects can take place in certain conditions. These changes are also due to the induced anisotropy of the internal stresses of the sample.

Consequently, the laws established for two minerals differing in composition, crystallographic structure and elastic characteristics are similar. This evidently is valid for other minerals as well since the studies of plagioclase, microcline and biotite have revealed that anisotropy of elastic characteristics does not remain constant with pressure change (fig. 1—4).

Biotite is characterized by increase of σ with pressure in the plane perpendicular to cleavage, that is plasticity of the mineral is growing whereas in the plane parallel to cleavage a decrease of σ is observed, implying increase of brittlenes. So, in biotite (fig. 4b)

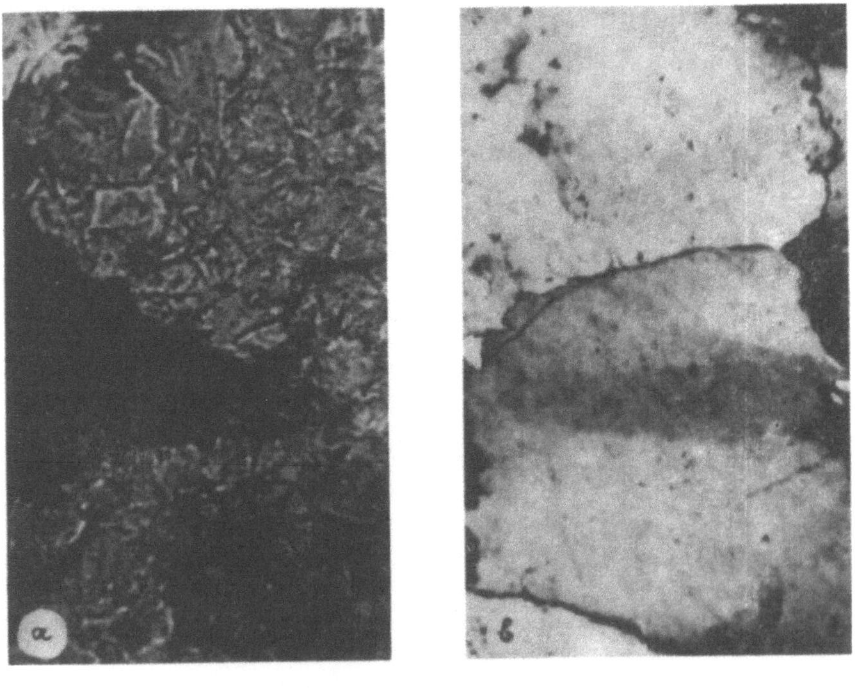

Fig. 8. Quartzite, thin section, parallel nicols, 72-fold magnification.
a — optical inhomogeneities in quartz grains at atmospheric pressure; b — quartz grains free of optical inhomogeneities after the effect of pressure of 5 kbar.

$\Delta\sigma$ between two orthogonal lines is 0.02 under atmospheric pressure, 0.18 to 3 kbar, and as large as 0.40 at 10 kbar. Further, anisotropy remains almost unchanged to 15 kbar.

The phenomena described for isolated crystals are also characteristic of the grains of similar minerals constituting the rock. This is confirmed by studying quartz crystals in quartzite [48]. A quartzite sample was examined at hydrostatic pressure to 15 kbar. In its grains optical inhomogeneities were found similar to those revealed in morion (fig. 8a). After the effect of more than 3 kbar no inhomogeneities were detected (fig. 8b). If the latter are related to the dislocational structure and characterized by internal remanent stresses [43] it is evident that both in individual quartz grains and in a single crystal under pressure the interaction of the applied external and internal remanent stresses may set in a motion of dislocations resulting in a nondislocational structure of individual grains. The grain boundaries may serve as inlets for dislocations. In a polycrystalline combination, hydrostatic pressure moving the dislocations may be lower than that in a single crystal, since the quartz grains are affected by additional anisotropic stresses caused by disorientation of the surrounding grains [48].

The complex processes in rocks may be investigated by studying the change of magnetic properties in rocks and minerals under high pressure. The magnetic characteristics are closely related to magneto-molecular and defect structures of rocks and minerals. Therefore, the behaviour of magnetic minerals in silicate matrix under hydrostatic loading yields information about the character of the change of internal remanent stresses and, hence, of the defect structure. The high pressure studies of remanent magnetization of rocks and minerals [49] confirm occurence of these phenomena in minerals and polycrystalline associations.

Finally, the analysis of our experimental results suggests the following pattern of the elastic parameters behaviour at hydrostatic pressure. To 1—2 kbar, the effect of macrodefects (porosity, intergranular contacts, fissures etc.) is being smoothed and elastic constants grow. Simultaneously, internal stresses within the sample volume are being redistributed. The pressure increase from 2 to 5 kbar yields a higher anisotropy of elastic properties of minerals due to mobility of the defects of crystal lattice. Under an optimum loading, the defects abandon the crystal. This perfects the crystal structure and, hence, improves the elastic parameters of the entire mineral association. At the same time in the pressure range under study, the system tends to perfection and a contribution of the forces of the atom-molecular interaction to the change of elastic parameters becomes greater [18, 54]. For this reason, the perfection of the mineral structure is accompanied by increase of elastic constants of minerals and the entire rock. In the rocks generally formed by anisotropic minerals a nonuniform accumulation of elastic energy is observed both in individual minerals and in the whole volume of the sample. This causes macrodeformations of minerals in the form of plastic transformation and brittle destruction and sometimes induces changes in elastic properties of certain minerals.

The pattern proposed may be applied to all the rocks composed of crystalline grains. However, quantitative changes of their elastic parameters, at varying pressure, depend on the original state and mineral composition of the rock.

ELASTIC PROPERTIES OF MINERALS, AS POTENTIAL REPRESENTATIVES OF THE UPPER MANTLE, AT HIGH PRESSURES

E. I. Bayuk

The problem of the composition of the upper mantle has been discussed for a long time. It is understandable that the processes, including the geodynamic ones, occurring at depth are determined by the composition and properties of the material involved. The information on the material of the Earth's interior can be obtained from the products of volcanic activity during which xenoliths of ultrabasic composition are ejected along with the melt. Investigation has been made of inclusions found in diamonds. Natural diamonds are known to be formed at a pressure exceeding 40 kbar, i. e. at a pressure corresponding to a depth below the Earth's crust. For this reason inclusions in diamonds can be regarded as messengers from the interior. Many investigators report a constant composition of inclusions. As a rule, they are composed of pyrope, olivine, diopside, enstatite and chrome-spinels. Other minerals are mentioned, but they are not typical [61].

Experimental study of *P-T* diagrammes of natural samples of kimberlite pipes established stability fields of various mineral asemblages [62]. Besides, experiments on fractional melting and crystallization permitted the creation of chemical and petrographic models of the upper mantle [63]. A predetermined chemical composition of a model can be obtained by various combinations of silicates, such as garnet, olivine and pyroxene. Pyrope, forsterite and diopside are most probable among the groups of the indicated minerals, because a high content of magnesium oxide is specified by modelling. Some investigators indicate percentage of each mineral in models. Diopside and pyrope are supposed to be present in the largest amounts.

Apart from studying the composition of the upper mantle material and the conditions of its existance, it is important to know its physical properties and to compare them with the results of crustal geophysical investigations. Data of explosion seismology (or deep seismic sounding in Russian terminology) suggest the presence of seismic anisotropy of the upper mantle. So, the model must necessarily account for the anisotropy of seismic velocities.

The aim of this work is to analyse the results obtained by the author from velocity measurements in monocrystals of minerals from the clinopyroxene, olivine and garnet groups under high pressures. A discussion of the results available in the literature is also presented.

Specimens 12.5 or 17.5 mm in diameter and 20 mm in length were prepared from crystals in accordance with their crystal structure. The axis of a specimen cylinder coincided within 1° with one of the main crystal symmetry directions. Specimens were tested in an apparatus of a cylinder-piston type at pressure to 20 kbar. Lead was used as a material exerting pressure on a specimen. Velocities of elastic waves were measured by an ultrasonic pulse method [64]. Ultrasonic waves were generated and detected by ceramic transducers (lead zirconate-titanate) having a natural frequency of 2—3 Mcps. Velocity measurements were made with an accuracy of 1—2%.

When studying elastic properties of an isostructural succession of clinopyroxene minerals, experiments were made to investigate the depencence of velocities on chemical composition and direction of measurement.

The principal crystallographic characteristics of a clinopyroxene structure are simple chains of SiO_4-tetrahedrons with a $[S_2O_6]$ radical, bonded by continuous structures of cation octahedrons extending in the same direction. The chains are elongated parallel to the crystal axis c. This is the direction in which the largest values of wave velocities are expected. Since bonding is little different in degree in clinopyroxene along crystal axes, velocities are approximately identical.

Directions for measuring velocities were selected in accordance with pyroxene crystal structure and with the results obtained from velocity measurements under atmospheric pressure. Under high pressure compressional velocity was measured in the direction [001], along the chains, and in [010], across them.

Fig. 1 shows variation of compressional velocity with pressure for spodumene, jadeite, diopside, aegirine and hedenbergite. Since the experiments were made on natural crystals having microfractures, the velocity-versus-pressure relationships appeared to be similar to those obtained from experiments on rocks, i. e. velocity rapidly incereased at low pressure. In order to determine the initial velocity for non-porous minerals, the linear part of the curve $v(P)$ at high pressure was extrapolated to atmospheric pressure. As seen from the curves, compressional velocity in all clinopyroxenes turned out to be higher in the direction [001] than in [010], the difference being 6—16%. The largest coefficient of anisotrpy was found in diopside, and the smallest in jadeite. A relatively high velocity anisotropy in clinopyroxenes seems to be responsible for different velocities in different directions in clinopyroxens or in other rocks containing oriented clinopyroxenes.

The data on the dependence of density and velocity on chemical composition of clinopyroxenes are listed in table 1. The table shows percentages of essential oxides for each

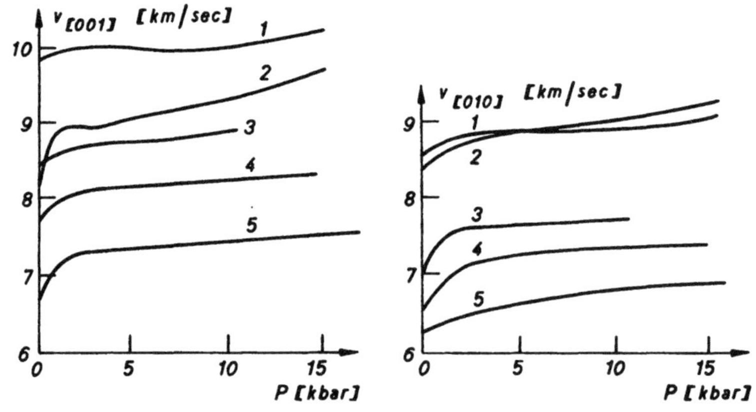

Fig. 1. Velocity of elastic waves vs. pressure for pyroxenes: a) [001], b) [010] for *1* — spodumen, *2* — jadeite, *3* — diopside, *4* — aegirine, *5* — hedenbergite.

Table 1. Chemical composition and physical properties of clinopyroxenes

Mineral	Chemical composition, %							$m = \frac{\overline{M}}{n}$	ϱ g/cm³	v_{av} km/sec
	SiO_2	Al_2O_3	FeO^+ Fe_2O_3	MgO	CaO	Na_2O	Li_2O			
Spodumene	64.5	27.4					8.1	18.6	3.11	9.2
Jadeite	59.4	25.2				15.1		20.2	3.24	8.5
Diopside	55.6			18.5	25.9			21.6	3.22	8.0
Algirine	52.0		34.6			13.4		23.1	3.55	7.5
Heden-bergite	48.4		29.2		22.2			24.8	3.41	6.8

mineral, mean atomic weight, $m = \frac{\overline{M}}{n}$ ($\overline{M}$ — molecular weight, n — number of atoms in molecule), density, ϱ, and average compressional velocity, v_{av} (for non-porous specimens).

The presence of Al and Mg cations increases velocity and that of Fe cations increases density. Velocity is inversely proportional to the mean atomic weight of a mineral. For an isostructural succession of clinopyroxenes this relationship is rather close, being expressed by the relation $v = -0.38m + 16.25$. Using this relation, we can redict velocities of those clinopyroxenes which do not possess crystals convenient for experimental study.

Olivine is an intermediate number of a isomorphous series from forsterite Mg_2SiO_4 to fayalite Fe_2SiO_4. It belongs to orthosilicates having isolated tetrahedral groups. The structure of olivine is characterized by closest hexagonal packing, the octahedral positions between the SiO_4-tetrahedrons being occupied by bivalent cations.

Velocity measurements in dunite which essentially consisted of oriented olivine (96 per cent) [65] showed the highest velocity in the direction [100], i. e. in the direction of the highest reticular density. Still higher velocity was obtained in a monocrystal of olivine having a density of 3.36 g/cm³ [66]. As seen from fig. 2, olivine-bearing rocks have a rather high coefficient of velocity anisotropy which at 10 kbar reaches 16.8%. Velocity anisotropy is still higher in an olivine monocrystal. Like clinopyroxenes, oriented olivine present in the upper mantle must cause seismic anisotropy in the subcrustal part of the Earth.

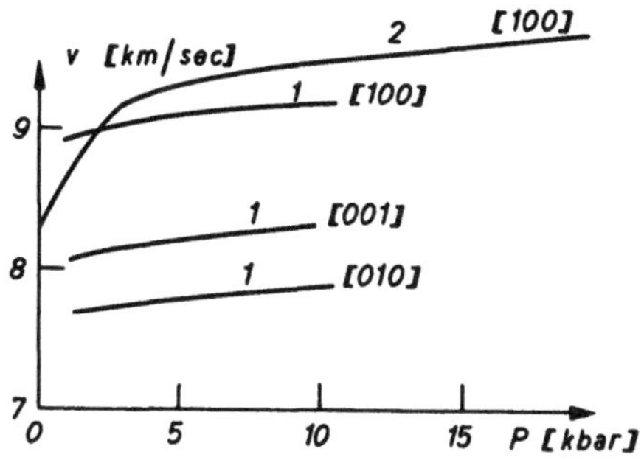

Fig. 2. Velocity of elastic waves in olivines vs. pressure *1* — data [65], *2* — data [66].

Experimental study of synthetic olivine specimens having various Mg and Fe oxide ratios revealed the dependence of physical properties (velocity, density, etc.) on chemical composition [67]. The largest MgO content in forsterite determined a high compressional velocity of the mineral (8.53 km/sec), whereas the maximum FeO content in fayalite produced a velocity of 6.63 km/sec. The equation expressing the velocity and mean atomic weight relationship for an isomorphous olivine series is $v = -0.2\,m + 14.97$.

Unlike clinopyroxenes and olivines, garnets do not manifest velocity anisotropy, being virtually isotropic. A slight difference in velocities was observed at atmospheric and low pressure, which seems to be a result of microfracturing. Natural garnets and garnetiferous rocks are usually strongly fractured. For this reason they have low velocities at atmospheric pressure, and rapidly increasing velocities at high pressure (Fig. 3).

Physical properties of garnets are dependent on the composition of cations (Table 2). One can see that, like in the case of clinopyroxenes and olivines, velocities in garnets increase as mean atomic weights decrease. The presence of Fe cations lowers velocity and raises density. The equation $v = 0.3\,m + 15.85$ is not quite dependable because the data on velocity measurements at high pressure in compositionally different garnets are insufficient. Velocity measurements for natural pyrope are not available. It is supposed that velocity in pyrope must be more than 9.5 km/sec.

Fig. 4 shows the curves of relationship between compressional velocities and mean atomic weights for garnets, olivines and clinopyroxenes. The slope of the curves is due to the

Table 2. Chemical and physical properties of garnets

Mineral	Chemical composition						$m = \frac{\overline{M}}{n}$	ϱ g/cm³	v_{av} km/sec	Reference
	SiO_2	Al_2O_3	FeO^+ Fe_2O_3	MgO	CaO	MnO				
Grossular	40.2	23.7			34.4		22.5	3.62	9.27	[68]
Almandine-pyrope							22.8	4.12	8.92	[69]
Same	37.6	22.4	33.2	5.07			23.8	4.16	8.53	[70]
Spessartite-almandine	36.4	20.6	19.4			22.3	24.9	4.25	8.43	[71]

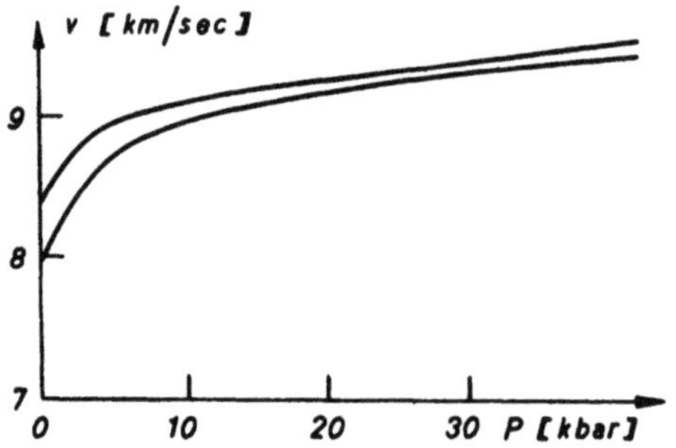

Fig. 3. Velocity of elastic waves in garnet as a function of pressure (data [69]).

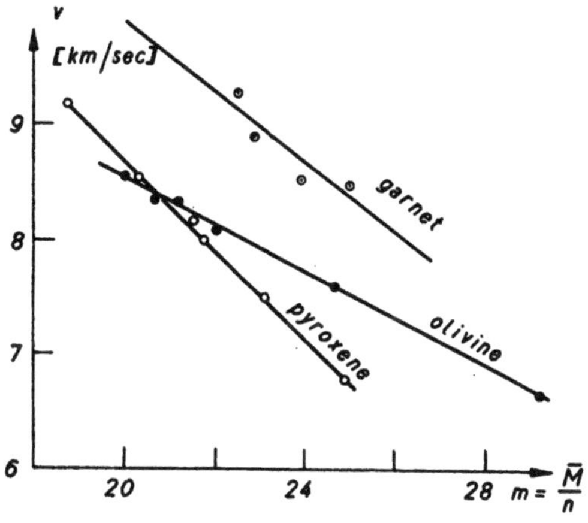

Fig. 4. Velocity of elastic waves vs. mean atomic weight: *1* — garnet, *2* — olivine, *3* — pyroxene.

crystalline structure of some mineral groups. The smallest slope was obtained for an isomorphous series of olivines having a simple formula. The largest slope was found for clinopyroxenes having a chain structure.

The minerals containing Mg cations are characterized by high elastic parameters. They are arranged in the following succession in order of decreasing velocity: pyrope, forsterite and diopside. Even in diopside compressional velocity is 8 km/sec. Estimating velocity variation with depth, namely increase with pressure and some decrease with temperature from experimental data [57], we cannot expect velocities to be much higher.

In conclusion, on the basis of the above mentioned experimental data, we were able to calculate possible values of compressional velocities in the proposed models of the upper mantle. The velocity for garnet pyrolite appeared to be 9 km/s and velocity anisotropy did not exceed 5 %.

For peridotite pyrolite these values were 8.5 km/s and 11 %, respectively. The model of peridotite pyrolite is seen to be more consistent with the requirement of seismic anisotropy.

RELATIONS OF THE VELOCITIES v_p AND v_s AND OF YOUNG'S MODULUS TO OTHER PETROPHYSICAL PROPERTIES FOR CLASTIC SEDIMENTARY ROCKS SUBJECTED TO A HYDROSTATIC PRESSURE

D. FREUND

Introduction

The velocities of elastic waves, v_p and v_s, and their behaviour under modified conditions (e. g. increased pressure) depend on the material-textural composition and the structure of the mineral matter. Consequently the velocity values in highly heterogeneous substances such as clastic sedimentary rocks are influenced by a great number of factors, the interaction of which has a very complex character.

Moreover, as the material heterogeneity increases, the determination of velocities presents more problems with respect to the measuring techniques, which especially applies to the determination of the velocity of shear waves, v_s.

Methods of determining v_s (and v_p) in sandstones under modified conditions

Measurements of v_s under increased pressure were for the first time undertaken by [72], where this quantity is measured, among other materials, also in sandstones. In long, thin cylindrical samples the lowest resonance frequency of torsional vibrations is measured (at up to 4 kbar and 150 °C). From this frequency and the sample length, v_s has to be determined.

D. S. Hughes and H. J. Jones describe in [73] a method of determining v_p and v_s, which is applied in [74], besides some other magmatic and metamorphous rocks, also on sandstones up to 5 kbar and +200 °C. This pulse method, which is designated as the PSP technique, is based on the fact that at the end face of a cylindrical sample a p-wave is converted into an s-wave, which is reflected under the critical angle, propagates through the cylinder, and is re-converted into a p-wave at the opposite end. Later is stated that this method is unsatisfactory for rocks [75].

Another method using the angle of total reflexion, as applied to sandstones at a pressure up to 0.2 kbar, is described by [76]. In this case a rotating sample is immersed in a liquid bath. By means of the rotation, the sample reaches positions where on the one hand compression waves are refracted at a plane parallel to the sample/liquid interface and, on the other hand, shear waves are refracted and totally reflected. This method of the critical angle in sandstones at pressures up to 0.7 kbar is used in [77].

These above-mentioned methods are only of minor importance as compared to the ultrasonic pulse technique described by [78, 79] for compressional waves and in [80] for shear waves. With the use of this method, extensive series of mainly magmatic and metamorphous rocks under pressure are investigated by many authors. According to this method the delay time of a plane compression or shear wave is determined in a cylindrical rock sample. In principle this pulse technique is most frequently and most successfully applied to rocks. Here any reference to the extensive literature shall be dispensed with. Only a small number of determinations of v_s under modified conditions in clastic sedimentary rocks are found in the literature.

Methods of investigation and investigated material

For determining v_p and v_s, a pulse technique is employed which operates with specially constructed ultrasonic transducers and permits a simultaneous determination of v_p and v_s waves. This fact proves extremely useful, above all in cases where further elastic constants of the investigated material shall be calculated from the velocities of the elastic waves. The employed high-pressure apparatus is explained in detail [9]; the ultrasonic measuring technique (electronic setup, special ultrasonic transducers), by [11].

The investigated samples were obtained from a bore put down in the north region of the GDR; they belong to the sedimentary Rotliegende (depth about 3000 m). Disks of the drill core having a height of 25—35mm are vertically cut into segments and used for the following analyses in addition to determining the v_p and v_s wave velocities:

1. Grain-size parameters
2. Density and porosity
3. Quantitative microsection analysis.

The grain size as determined by means of sieve analysis provides, among other things, the median. This value represents that grain size in mm which is smaller and greater, respectively, than 50 wt % of the sample material. The quantity ranges from 0.035 to 0.15 mm.

Macroscopic identification, microscopic sampling, and the granulometric curve derived from the sieve analysis provide the following information on the composition of the material.

Rocks	No. of samples
Siltstone	2
Fine sandstone	11
Medium-grained	6
Coarse sandstone	2

The density and porosity values as determined by the method generally applied in petrophysics range from 2.37 to 2.69 g/cm^3 (volume density) and from 0 to 11% (total porosity), respectively.

In addition to the exact rock identification, a quantitative determination of the constituents can be carried out by means of integration of a microsection of each sample. For a more detailed characterization of the elastic behaviour of the investigated clastic sediments, a distinction is drawn first of all between:

1. Clasts

These are more or less rounded sedimentary particles having minimum diameters up to 0.02 mm and

2. Binding material (or cement), which includes all particles smaller than 0.02 mm as well as the following types of binding agents:
a) clayey-sericitic
b) carbonatic
c) anhydritic
d) silicatic.

The percentage of clasts amounts to 50 to 90% of the investigated samples, that of clayey-sericitic cement ranges from 2 to 45%. Among the quantitatively inferior cement types, the carbonatic one occurs most frequently, ranging from 1 to 20%, whereas the silicatic and anhydritic cements amount to at the most 3%.

Results and interpretation

In the cylindrical samples of clastic sedimentary rocks, v_p and v_s have been measured up to maximum pressures of 5 kbar. Most of the air dry samples prepared normally to the plane of lamination were hydrostatically loaded by pressures up to 3 kbar.

The behaviour of v_p and v_s

The typical behaviour of v_p, i. e. its steep increase up to about 1.5 kbar, followed by a linear, gradual increase up to 3 and 5 kbar, respectively, is observed for v_s as well.

Up to 3 kbar the variations of v_p and v_s reach different values. While v_p changes by 22 to 115%, the increase in v_s amounts to 19 to 70% as compared to the initial velocities at normal pressure.

The comparison of the initial values $v_{p;s(0)}$ with the per cent variations $\Delta v_{p;s}/v_{p;s(0)}$ as functions of density and porosity does not reveal any oberservable dependence; neither the values of the total porosity nor those of the volume density have a distinct effect on

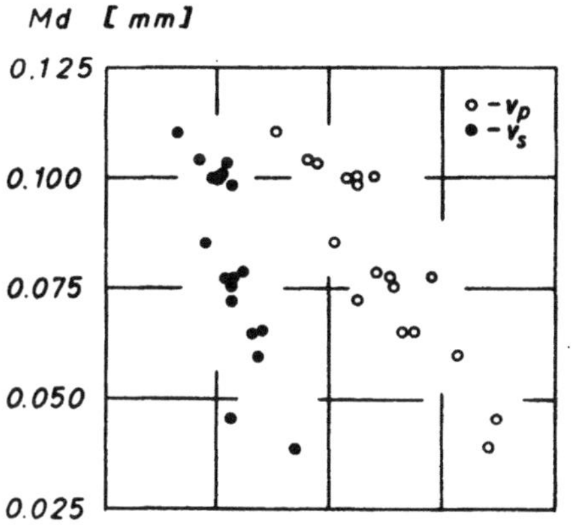

Fig. 1. Velocities v_p and v_s under normal conditions vs. the "median" grain-size parameter.

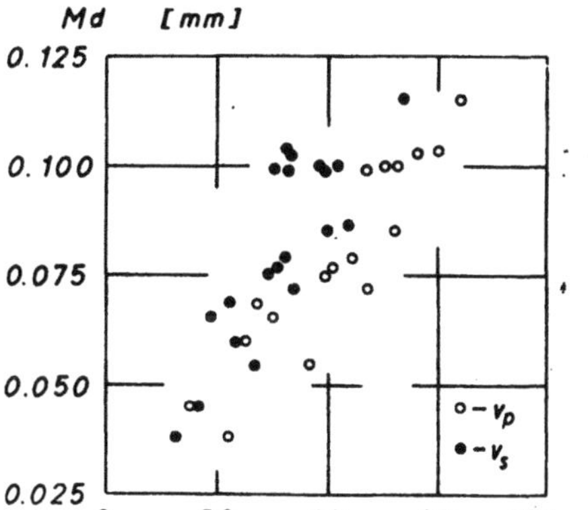

Fig. 2. Per cent variation of sound velocities at pressures up to 3 kbar vs. the median.

$v_{p;s(0)}$ and $\Delta v_{p;s}/v_{p;s(0)}$, which rather depend solely on the crack porosity. But this only accounts for a very small portion of the total porosity.

A clear relationship results from the "median" grainsize parameter (mm) for both the initial velocities $v_{p;s(0)}$ and the per cent variations of the two sound velocities up to 3 kbar (cf. Figs. 1 and 2).

The parameter of acoustic resistance, also called acoustic inertia, $v_p \cdot \varrho$ [km/sec cm³], shows an equally distinct dependence in that it increases with decreasing median values (see Fig. 3).

A dependence of the variation $\Delta v_{p;s}/v_{p;s(0)}$ on the values of the initial sound velocity, $v_{p;s(0)}$, is generally expected, but in Fig. 4 it shows up with surprising clearness, the more so if the highly different petrophysical parameters of the starting material, such as the porosity, the density, and the grain size, are taken into consideration.

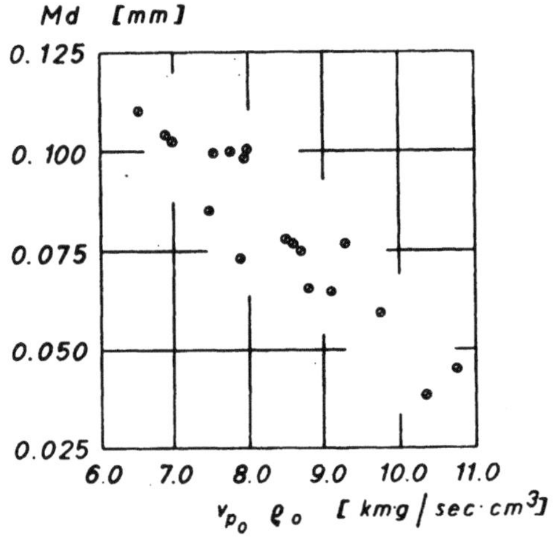

Fig. 3. Acoustic resistance at normal pressure vs. the median.

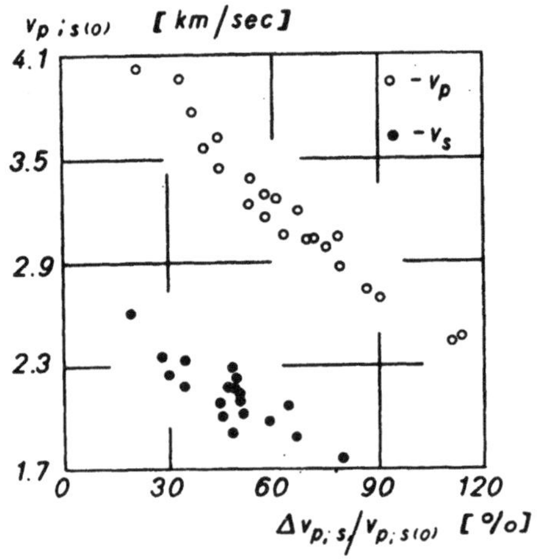

Fig. 4. $\Delta v_{p;s}/v_{p;s(0)}$ vs. the initial velocities $v_{p;s(0)}$.

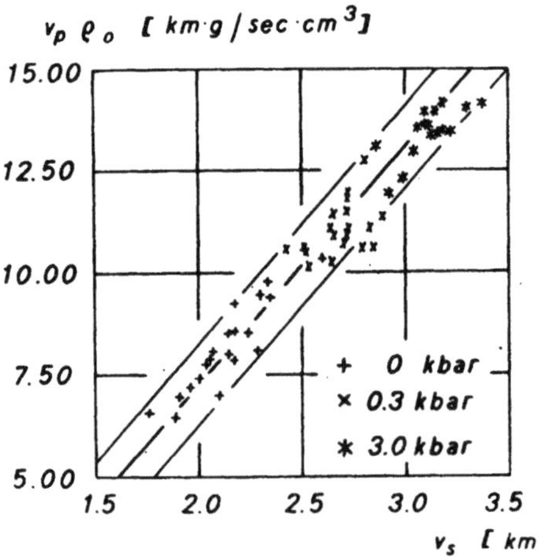

Fig. 5. Linear regression between $v_p\varrho_0$ and v_s at various pressures for v_p and v_s.

In Figure 5, the v_p value at 0; 0.3; and 3 kbar, multiplied by the density ϱ_0, is plotted vs. the shear-wave velocity (likewise for 0; 0.3; and 3 kbar). One obtains a linear dependence:

$$v_s(P) = [0.175 v_p(P) \cdot \varrho(P_0) + 0.75] \pm 0.15 .$$

According to this regressions equation, the corresponding values of $v_s(P)$ for the investigated samples can be calculated from the density ϱ_0 (volume density) and a velocity of the P wave, $v_p(P)$, measured at the pressure P.

Behaviour of Young's modulus E

Young's modulus E can be calculated from the values of the elastic velocities v_p and v_s (together with the density ϱ). Under pressure this parameter shows a behaviour analogous to v_p and v_s: a rapid increase in the pressure range up to about 1.5 kbar and a slow linear increase towards higher pressure values.

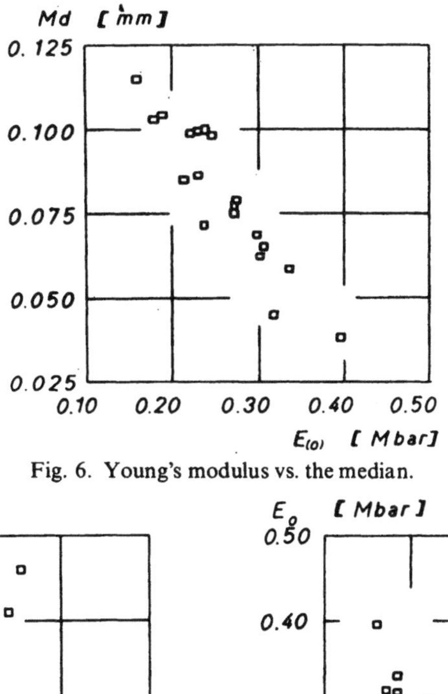

Fig. 6. Young's modulus vs. the median.

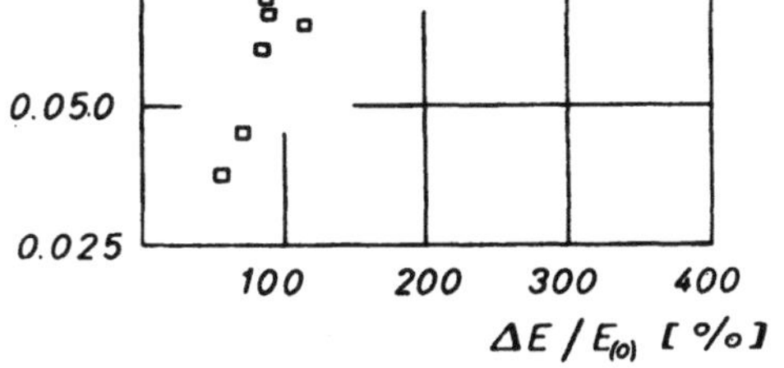

Fig. 7. Relations between the median and the per cent variation of Young's modulus, E, of the investigated clastic sedimentary rocks.

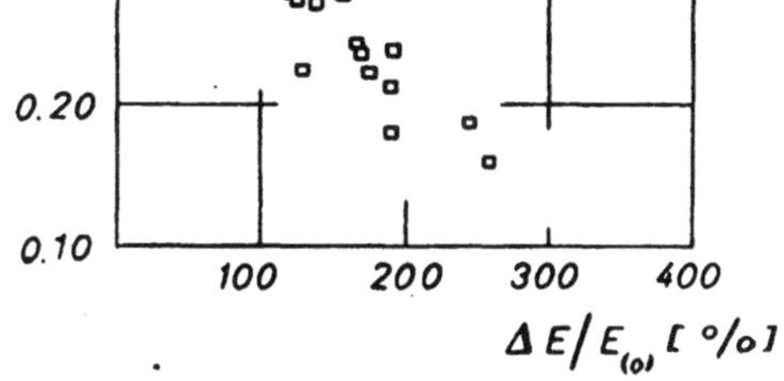

Fig. 8. Per cent variation of Young's modulus at pressures up to 3 kbar vs. Young's modulus at normal pressure.

The variation of $\Delta E/E_0$, however, reaches much higher values. For pressures up to 3 kbar it ranges from 60 to 260%.

Figures 6, 7, and 8 show relationships similar to those described by Figures 1, 2, and 4. Here merely the values of Young's modulus (E_0, $\Delta E/E_0$) occur instead of the v_p and v_s data.

From the behaviour of E_0 and $\Delta E/E_0$ as referred to the total porosity (Figs. 9 and 10), it is possible to derive an additional subdivision into sandstones with contents of carbonatic cements being higher or lower than 7.5%, respectively.

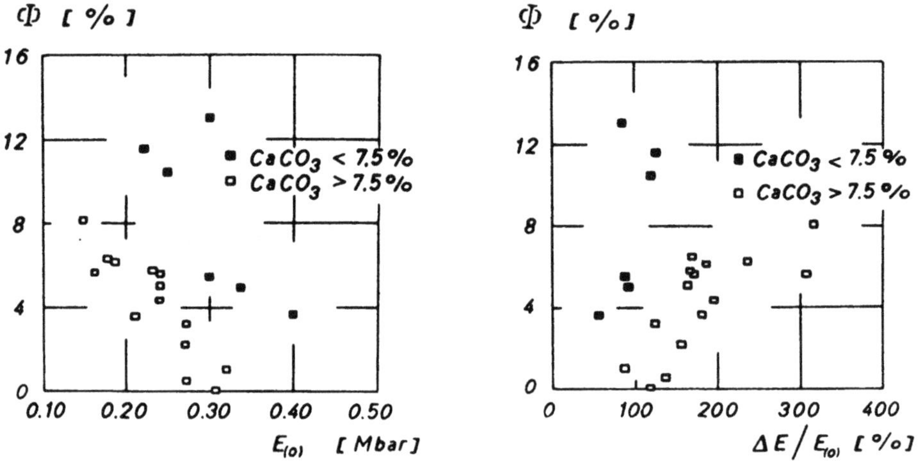

Fig. 9 and 10. Initial values and variations of Young's modulus vs. the total porosity Φ for different contents of carbonatic cement.

The crack porosity as the most significant characteristic of the behaviour of v_p, v_s, and E

As already mentioned above, the magnitude of crack porosity can be considered the cause of the initial soundvelocities $v_{p;s(0)}$, but expecially of the degree of per cent variation of the sound velocities in lower pressure ranges up to 2—3 kbar. The same also holds for the quantities E_0 and $\Delta E/E_0$.

Micro-cracks can hardly be detected under the microscope, but their quantitative measurement presents even greater problems unless the method of fluorescent X-ray analysis is used, as for instance in granites [81].

Micro-cracks may occur as intercrystalline cracks on grain contacts or as intracrystalline ones in the form of, say, cleavage and translation cracks. The two latter possibilities should occur in feldspar clasts as well as in ahydritic or carbonatic cements. The feldspar portion in the investigated samples is however very small, likewise the small portion of the anhydritic cement does not play any role. On the other hand the relatively high carbonate content would suggest a high amount of variation of v_p and v_s in the covered pressure range. But also in this case it was not possible to observe a significant effect. Consequently the major portion of crack porosity, which is infinitely small as compared to the total pore volume but decisive for the behaviour of v_p, v_s, and E, must occur in the intercrystalline region.

An increased crack porosity leads to lower initial sound velocities $v_{p;s(0)}$ and to high amounts of the percent variation $\Delta v_{p;s}/v_{p;s(0)}$ over the covered pressure range. The median shows the same behaviour, as it can be observed in Figs. 1 and 2. It can be shown that the crack porosity is an intercrystalline phenomenon, that its degree increases with increasing grain size, and that it occurs predominantly on grain boundaries of larger sedimentary particles. Analogous statements apply to the behaviour of Young's elastic modulus (cf. Figs. 6 and 7).

I am deeply indebted to Dr. T. S. Lebedev, Director of the High-Pressure Laboratory of the Geophysical Institute of the Ukrainian Academy of Sciences, Kiev, who kindly permitted that the measurements were carried out at this laboratory.

My particular gratitude is due to V. A. Korchin, V. T. Vasilyaka, and V. I. Shapoval for carrying out the measurements and for stimulating discussions.

DYNAMICS OF CHANGES IN ELASTIC PROPERTIES OF ROCKS UNDER VARYING *PT*-CONDITIONS OF THE EARTH'S CRUST

T. S. LEBEDEV, V. A. KORCHIN

Introduction

The results of the high pressure and high temperature studies of rocks and minerals provide solutions for important problems related to the nature and regularities of the change of elastic properties within the Earth's interior. These results are widely applied to the theoretical and experimental modelling of different physico-mechanical, physico-chemical, tectonophysical and other processes and phenomena at great depth. Along with evidence from the ground surface studies of different geophysical fields, the data of the said experiments make possible a more accurate prediction of the state and composition of mineral matter and the deep-crustal structure [9, 12, 16, 42, 57, 58, 82, 85, 87, 92—94, 122, 123, 125 etc.].

But however large is the amount of the experimental data, they are not sufficient for a complex analysis of the distribution of the mineral masses within the interior if the experiments provide the high pressure only. The data on the mineral matter behavior under a simultaneous effect pressure (*P*) and high temperature (*T*) are indispensable. Because of many technical and methodical difficulties encountered in of this kind of experiments they are still very rare. The number of the rock samples that have been studied in the world since 1937 by using this technique is about 100 [2, 6, 7, 9, 10, 12, 14, 16, 58, 73—75, 83, 84, 86, 88, 90, 95—97, 103, 105, 111, 113, 115, 124].

This section presents the results of experimental studies of longitudinal (V_p) and transversal (V_s) waves and elastic parameters in rocks under varying *P* and *T* effects. On the basis of information provided by the data on the elastic characteristics, variation in the mineral matter under high hydrostatic and quasihydrostatic pressures and high temperatures, and, eventually, by the results of the simultaneous *P* and *T* modelling, a possible is presented for the upper crustal rocks for which different versions of pressure and temperature change with depth are pregiven.

The studies are completed at the Department of Physical Properties of the Earth's Matter, Institute of Geophysics, Academy of Sciences of Ukr. SSR, Kiev. The high pressure apparatus included devices providing hydrostatic and quasihydrostatic pressure to 6 kbars and temperature to 320—350 °C [6, 9, 13, 116—118 etc.]. The experiments were performed by the method of programmed modelling of three thermodynamic regimes dominating within the deep crustal zones under the Ukrainian Precambrian Shield [12, 15, 119, 120]. Elastic parameters were computed from simultaneous V_p and V_s measurements in rock samples [11, 12, 14, 16, 95, 105, 114, 115 etc.] and density data.

Studying elasticity moduli and coefficients by dynamic methods is fairly important for

Table 1. Mineral composition (volume %) and physical properties of certain rocks from the Ukrainian shield under atmospheric pressure and room temperature

No.	Rocks and sampling regions	SiO_2 cont. (weight %)	Pl	Potash feld-spar	Q	Px	Bi	Hb	Gr	Acces. and ore minerals
	Acid rocks									
1.	Quartzite, Ovruch	97.35	—	—	100	—	—	—	—	—
2.	Charnokite I, Chernovtsy	72.85	34.3	21.8	38.3	3.4	—	—	—	0.8
3.	Zhitomirsky granite I, Boguşlav	72.84	35.0	23.0	23.0	—	—	9.6	—	2.0
4.	Kirovogradsky granite, Kirovograd	72.29	28.4	36.2	24.5	—	—	9.5	—	1.4
5.	Porphyry granite I. Pugachyovka	71.89	35.2	27.3	29.4	—	—	5.2	—	2.9
6.	Zhitomirsky granite II, Zhitomir	71.62	47.4	23.0	23.3	—	—	6.3	—	—
7.	Charnokite II, Khoschevatoye	71.24	52.1	—	27.9	16.1	—	—	—	3.9
8.	Rapakivi granite I, Penizevichi	71.24	6.1	60.7	17.8	—	3.9	1.7	—	3.3
9.	Porphyry granite II, Bekhy	70.71	29.2	21.7	34.8	—	—	8.9	—	5.6
10.	Rapakivi granite II, Korsun-Shevchenkovsky	70.13	16.0	63.8	11.5	—	4.6	2.8	—	1.2
11.	Chudnovo-Berdichevsky granite, Zhezhelevo	64.97	30.3	17.2	28.7	-	-	13.4	10.4	—
12.	Klesovsky granite, Klesovo	64.65	56.0	7.0	12.6	—	6.8	8.0	—	4.4
	Intermediate rocks									
13.	Charnokite III, Zavalye	64.16	62.5	—	19.8	15.4	—	—	—	1.1
14.	Sobite, Kalinino	62.87	55.9	8.0	5.7	5.7	—	11.8	—	8.1
15.	Granodiorite, Shepetovka	61.96	49.4	3.1	21.6	—	—	19.3	—	6.6
16.	Diorite I, Trigur'ye	61.95	45.7	11.4	14.1	—	9.6	16.0	—	3.2
17.	Osnitsky granite, Klesovo	59.43	66.7	7.5	7,5	—	3.4	8.3	—	6.6
18.	Klesovsky porphyrite, Tomashgorod	54.12	49.7	2.3	10.8	—	18.1	11.6	—	7.5
19.	Diorite II, Rakitnoye	52.63	52.7	—	7.8	—	19.7	17.5	—	2.3
20.	Diorite III, Viry	52.80	41.5	3.8	5.1	—	16.1	14.0	—	4.1
	Basic rocks									
21.	Labradorite I, Kamenka	54.94	61.7	12.0	14.7	3.1	1.1	—	—	3.0
22.	Labradorite II, Slipchitsy	52.97	84.0	—	2.9	4.6	—	—	—	6.2
23.	Gabbro-norite, Golovino	52.27	67.7	—	—	29.3	—	—	—	3.0
24.	Diabase, Klesovo	46.31	61.6	—	—	25.8	—	—	—	5.5
25.	Gneiss, Zavalye	46.00	39.5	—	7.4	14.6	14.1	—	5.3	4.9
26.	Pyroxenite, Fastov	44.18	—	—	—	43.5	—	—	—	12.4
27.	Eclogite-like rock I, Khoschevatoye	59.05	28.6	—	42.6	—	—	—	26.0	2.8
28.	Eclogite-like rock II, Zavalye	46.60	26.6	—	3.4	28.8	12.1	—	22.6	4.1

Secondary minerals	δ g/cm³	ϱ g/cm³	n %	V_p km/sec	V_s km/sec	V_p/V_s	E Mbar	G Mbar	K Mbar	β Mbar⁻¹	σ	λ Mbar
—	2.650	2.652	≈0	5.91	3.93	1.50	0.90	0.41	0.38	2.63	0.10	0.11
1.4	2.689	2.649	1.5	5.86	3.52	1.66	0.80	0.33	0.47	2.12	0.22	0.25
—	2.669	2.643	1.4	5.39	3.17	1.70	0.66	0.26	0.41	2.42	0.23	0.25
—	2.644	2.645	≈0	5.44	3.21	1.69	0.67	0.27	0.42	2.38	0.23	0.24
—	2.660	2.629	1.0	5.88	3.66	1.61	0.77	0,32	0.44	2.43	0.19	0.20
—	2.655	2.639	0.6	5.53	3.30	1.67	0.70	0.28	0.42	2.35	0.22	0.23
—	2.770	2.756	0.5	6.35	3.82	1.66	0.98	0.40	0.58	1.74	0.22	0.31
6.5	2.695	2.660	1.3	6.04	3.40	1.78	0.78	0.30	0.56	1.82	0.27	0.35
—	2.675	2.647	1.1	5.69	3.21	1.77	0.69	0.27	0.49	2.03	0.27	0.31
—	2.645	2.636	0.3	5.24	3.55	1.56	0.68	0.33	0.33	3.04	0.15	0.13
—	2.799	2.777	0.8	5.56	3.28	1.70	0.74	0.30	0.46	2.22	0.23	0.26
5.2	2.738	2.737	≈0	6.03	3.28	1.84	0.76	0.29	0.60	1.66	0.29	0.41
1.2	2.911	2.865	1.6	6.24	3.51	1.78	0.90	0.35	0.64	1.55	0.27	0.41
3.8	2.770	2.733	1.4	5.63	3.30	1.71	0.74	0.30	0.47	2.13	0.24	0.27
—	2.754	2.756	≈0	5.58	3.38	1.65	0.76	0.31	0.44	2.28	0.21	0.23
—	2.752	2.744	0.3	5.53	3.22	1.72	0.71	0.28	0.46	2.17	0.24	0.27
—	2.752	2.719	1.3	5.69	3.07	1.85	0.66	0.26	0.54	1.86	0.29	0.37
—	2.753	2.759	≈0	5.74	3.45	1.66	0.80	0.33	0.47	2.12	0.22	0.25
—	2.866	2.860	0.2	5.75	3.40	1.69	0.81	0.33	0.50	2.00	0.23	0.28
9.4	2.903	2.861	1.5	5.34	3.24	1.65	0.73	0.33	0.42	2.41	0.21	0.22
4.4	2.817	2.817	≈0	6.18	3.38	1.82	0.83	0.32	0.65	1.54	0.29	0.49
2.3	2.767	2.761	0.2	6.40	3.59	1.78	0.90	0.36	0.66	1.52	0.27	0.42
—	2.977	2.977	≈0	6.91	3.75	1.84	1.08	0.42	0.86	1.16	0.29	0.58
7.1	3.035	3.035	≈0	6.07	3.63	1.67	0.99	0.40	0.58	1.72	0.22	0.32
14.2	2.744	2.736	0.3	5.51	3.11	1.77	0.67	0.26	0.48	2.09	0.27	0.3
44,1	3.925	3.273	0.7	6.55	3.76	1.74	1.16	0.46	0.79	1.27	0.25	0.48
—	3.192	3.151	1.3	6.57	3.85	1.71	1.16	0.47	0.74	1.36	0.24	0.43
2.4	3.192	3.172	0.6	6.77	3.72	1.82	1.13	0.44	0.87	1.15	0.28	0.58

understanding of cohesive forces in solids since the theses parameters reflect quantitavely the character of the interaction between micro- and macrostructural units within the solids. The regularities in the behaviour of the elastic characteristics then can always be useful for explaining the essential changes in the state of the material. Acoustical methods based on the elastic velocity measurements provide the means for studying the properties of the solids under low and high temperatures, within the high and super-high pressure domain, as well as dynamics of different physico-chemical transformations of the solid etc. On considering the material as a continuous medium, as the theory of elasticity does, elastic characteristics of the isotropic body can be expressed in therms of elastic velocities and density [91, 99]. The state of the solid under various elastic deformations can be expressed by the Young (E), shear (G) and compression (K) moduli, Poisson's ratio (σ), and Lamé constants (λ and μ).

Under uniform pressure applied to the multicomponent polycrystalline mineral matter (usually quasiisotropic), the internal stresses of a complicated character originate within the rock sample to cause the change in its volume and shape. These phenomena are the best described by compression and shear moduli, respectively. Simultaneously, under uniform pressure, the internal stresses in rocks bring about development of local axial and lateral deformations necessarily causing microstructural changes. The value of axial deformation in the rock sample depends, mainly, on its immediate resistance to the effect of the hydrostatic pressure forces and is characterized by the Young modulus. The Poisson's ratio is defined by the possibility of the rock to "transmit" the effect of the external stresses along the directions differing from those of the hydrostatic compression forces. This value is the source of information about the presence and character of the effect of the internal stresses in samples. It difines qualitatively the intensity of the development of elasticity and plasticity processes in the deformed samples. The information derived from the elastic parameters of rocks shows the importance of their studies in different thermobaric conditions.

Elastic properties of rocks under hydrostatic pressures and room temperature

Table 1 shows the constituent minerals and some physical characteristics for the rocks of acid, intermediate and basic composition sampled from the Ukrainian Shield. Fig. 1—3 presents the results of the longitudinal (V_p) and transversal (V_s) ultrasonic velocity measurements in these rocks under high hydrostatic pressures and room temperature. Omitting the details in description of the $V_{p,s} = f(P)$ relations for each sample we shall establish the general regularities for the three groups of samples. V_p and V_s grow with increase of basicity. The lowest mean velocities under atmospheric pressure are characteristic of acid rocks. ($V_{p001} = 5.25 \div 6.00$ km/sec; $V_{s001} = 3.15 \div 3.5$ km/sec) (fig. 1). The rocks of intermediate composition have the velocities of $5.5 \div 6.35$ and $3.10 \div 3.80$ km/sec (fig. 2). The highest velocity values go with the basic rocks: $V_{p001} = 6.0 \div 6.8$ and $V_{s001} = 3.15 \div 3.85$ km/sec (fig. 3). Under hydrostatic pressure to 5 kbar the velocity distribution for the three groups of rocks is essentially the same. So, $V_{p5} = 6.6 \div 6.8$ and $V_{s5} = 3.4 \div 3.7$ km/sec for acid rocks, $V_{p5} = 6.20 \div 7.00$ and $V_{s5} = 3.65 \div 4.00$ km/sec for intermediate, and $V_{p5} = 6.75 \div 8.10$ and $V_{s5} = 3.7 \div 4.2$ km/sec for basic rocks.

However, judging by the curves the velocity values scatter considerably under atmospheric and high pressures even for one variety of rocks. This is due to mineral composition,

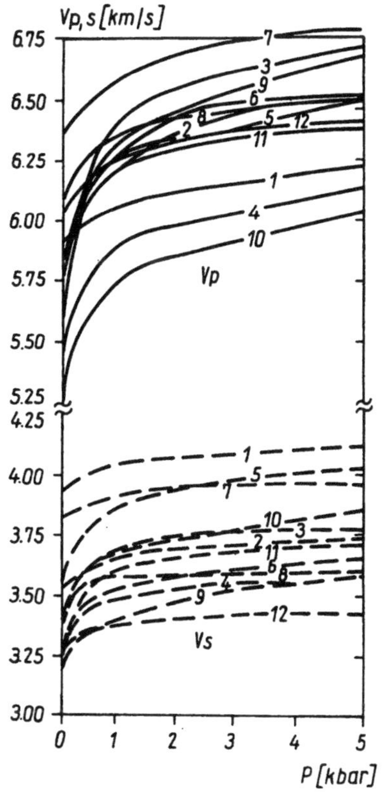

Fig. 1. Longitudinal (V_p) and transversal (V_s) velocities against hydrostatic pressure plots for acid rocks. *1* — quartzite; *2, 7* — charnokites I. II; *3, 6* — Zhitomirsky granites I, II; *4* — Kirovogradsky granite; *5, 9* — porphyry granites I, II; *8, 10* — rapakivi granites I, II; *11* — Chudnovo-Berdichevsky granite; *12* — Klesovsky granite.

texture-structural features and porosity [9, 12, 16, 100—103, 112, etc.]. In acid rocks, the highest velocities under atmospheric pressure are characteristic of the coarse-grained varieties with low porosity (fig. 1, s. *6, 11*) or of the rocks with a higher quartz content (fig. 1, s. *1*). With pressure growing to maximum values, the highest velocities are found in equigranular granite and charnokite as well as in rocks with a higher content of plagioclase (30 ÷ 40 %) and moderate content of quartz (20 ÷ 25 %). In coarsegrained granite with inconsiderable porosity the $V_{p,s} = f(P)$ values increase with small gradients.

The higher velocities in the deposits of intermediate composition are due to the presence of pyroxene and hornblende. With growing pressure the maximum changes in velocities have proved to be assigned in the rocks with a higher content of plagioclase and biotite (fig. 2, s. *5*). Velocity increase is the smallest in the samples rich in quartz (fig. 2, s. *1, 3, 6*). In the rocks of intermediate composition V_p and V_s change in the initial pressure range more smoothly, the linear section of the $V_{p,s} = f(P)$ curve appearing somewhat later. This is due to a small content of quartz and abundance of plagioclase and biotite plasticizing under pressure the entire rock.

Basic rocks acquire high values of V_p and V_s with a large scatter in any range of pressure (fig. 3). This is due, in the first place, to the presence of the constituent

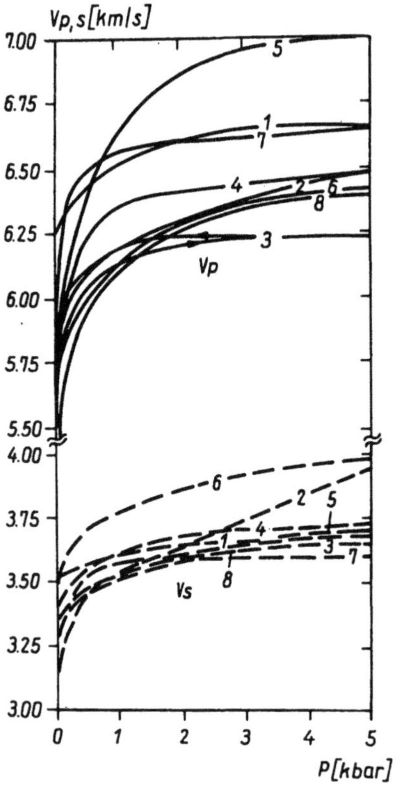

Fig. 2. Longitudinal (V_p) and transversal (V_s) velocities against hydrostatic pressure plots for intermediate rocks.
1 — charnokite III; *2* — sobite; *3* — granodiorite; *4, 7, 8* — diorites I, II, III; *5* — Osnitsky granite; *6* — Klesovsky porphyrite.

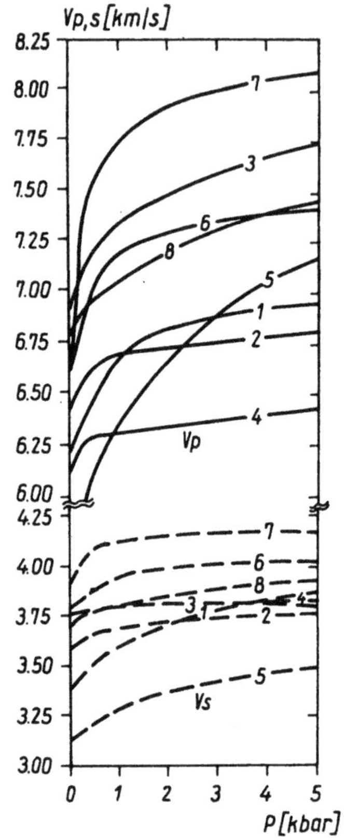

Fig. 3. Longitudinal (V_p) and transversal (V_s) velocities against hydrostatic pressure plots for basic rocks.
1, 2 — labradorites I, II; *3* — gabbro norite; *4* — diabase; *5* — gneiss Amf-Px-Pl; *6* — pyroxenite; *7, 8* — eclogite-like rocks I, II.

minerals with higher velocities (garnet, hornblende). As to the wide spectra of the observed velocities, these are caused by a variety of the constituent minerals. The latter explains the low $V_{p,s}$-values in amphybole pyroxene plagioclase gneiss (fig. 3, s. *5*). Here, the higher content of secondary minerals, a small amount of plagioclase and, possibly, a layered texture of the rock are also responsible. A rapid growth of longitudinal velocity under pressure is also explained by the original "friability" of the rock (due to its complex composition) and the presence of such minerals as quartz, hornblende, pyroxene and garnet. The tendency of basic rocks to acquiring the monomineralic composition causes an increase of velocities (fig. 3, s. *3, 6*). However, the monomineral labradorite (fig. 3, s. *1, 2*) shows no high velocities, since, as it microscopically found in its sections, it consists of widely disoriented large grains of labrador with considerably deformed borders. The highest velocities under high pressure are recorded for the eclogite-like rock containing the

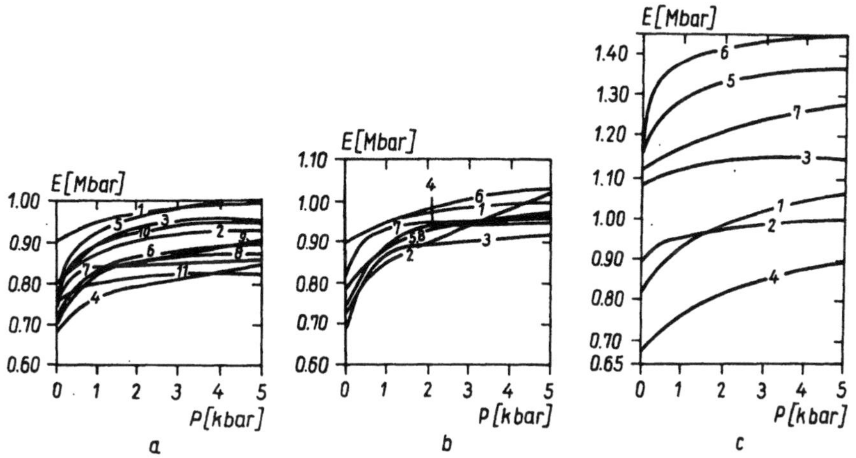

Fig. 4. Young modulus against hydrostatic pressure plot.
a — acid rocks: *1* — quartzite; *2* — charnokite I; *3, 6* — Zhitomirsky granites I, II; *4* — Kirovogradsky granite; *5, 8* — porphyry granites I, II; *7, 9* — rapakivi granites I, II; *10* — Chudnovo-Berdichevsky granite; *11* — Klesovsky granite. b — intermediate rocks: same legend as in fig. 2. c — basic rocks: *1, 2* — labradorites I, II; *3* — gabbro norite; *4* — gneiss Amf-Px-Pl; *5* — pyroxenite; *6, 7* — eclogite-like rocks I, II.

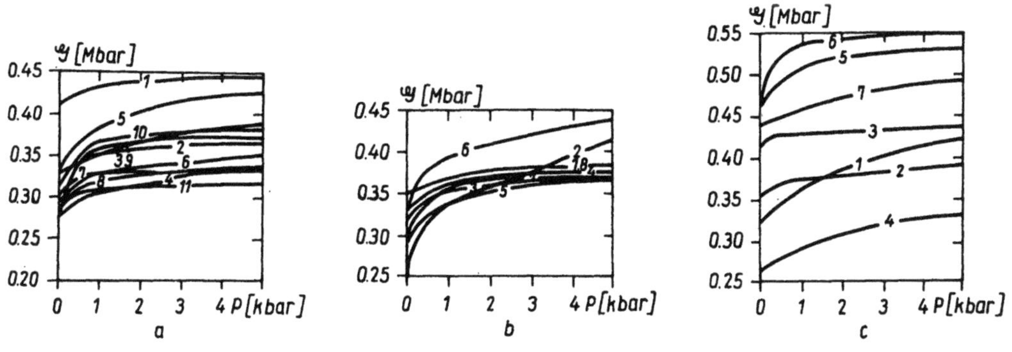

Fig. 5. Shear modulus against hydrostatic pressure plot.
a — acid rocks; b — intermediate; c — basic rocks. Same legend as in fig. 4a, b, c.

maximum amount of garnet (26%) implanted in quartz (42.6%). The latter is an elastic material producing the high quality acoustical contact between the minerals. An interaction between quartz and plagioclase filling in the major volume of the sample is responsible for a gradual increase of velocities with pressure.

Fig. 4—7 show the changes of elastic parameters with pressure. The behaviour of the Young and shear moduli (fig. 4, 5), compressibility (β; fig. 6) and Poisson's ratio (fig. 7) is demonstrated. With pressure growing from atmospheric to 2 kbar the moduli axial (E) and lateral (G) deformations are increasing by the curvilinear law with high gradients as are velocities. The further growth of pressure produces a linear change of E and G with smaller gradients. Alike to the modulus, the velocity values grow with basicity of rocks. Acid rocks under atmospheric pressure are characterized by a wide range of E-values

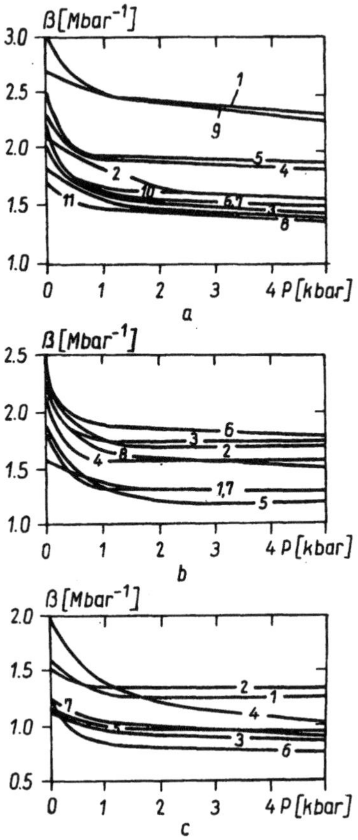

Fig. 6. Compressibility variation with hydrostatic pressure.
a — acid rocks, b — intermediate rocks, c — basic rocks. Same legend as in fig. 4a, b, c.

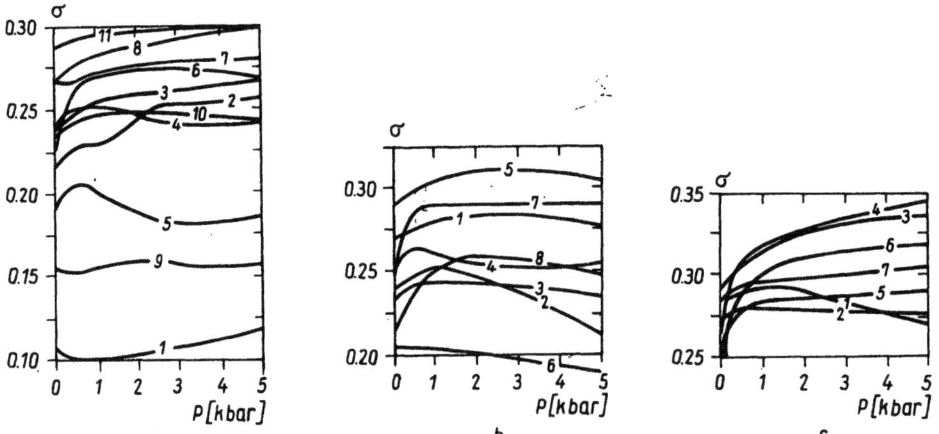

Fig. 7. Poisson's ratio against hydrostatic pressure plot.
a — acid rocks, b — intermediate rocks, c — basic rocks. Same legend as in fig. 4a, b, c.

($\Delta E_{001} = 0.90—0.45 = 0.45$ Mbar). To 5 kbar, this range narrows to 0.20 Mbar. Similarly, for intermediate rocks, ΔE_{001} is equal to 0.35 and ΔE_5 to 0.15 Mbar. In basic rocks, the range of E_0-values keeps this wide under pressure ($\Delta E_{001} = 0.5$; $\Delta E_5 = 0.55$ Mbar). In acid rocks, the coefficient of shear is less sensitive to the pressure effect ($\Delta G_{001} = 0.2$; $\Delta G_5 = 0.15$). In intermediate rocks, ΔG_{001} and ΔG_5 are 0.15 and 0.1, respectively, in basic rocks 0.2 and 0.22 Mbar.

Compressibility (β) change with pressure in the rocks under study has somewhat different regularities (fig. 6). At atmospheric pressure, the behaviour of the compressiblity value depends on basicity of the rock whereas under high pressure this does not hold. In acid rocks, $\Delta\beta_{001} = 3.6—1.6 = 2.0$ and $\Delta\beta_5 = 2.3—1.1 = 1.2$; in intermediate rocks $\Delta\beta_{001} = 2.4 - 1.6 = 0.8$, $\Delta\beta_5 = 1.8 - 1.2 = 0.6$; in basic rocks, $\Delta\beta_{001} = 2.0—1.2 = 0.8$; $\Delta\beta_5 = 1.3—0.7 = 0.6$ Mbar^{-1}. As is seen, basic rocks have the lowest values of compressibility. This is due to their texture-structural features and mineral composition. Our experiments with minerals have shown that quartz has the highest compressibility. This is confirmed by studying quartzite (fig. 6a, s. *1*). Hence, the growth of the amount of quartz should increase compressibility of the rock. The same effect is produced by K-feldspar. For instance, samples *4*, *5*, *9* (fig. 6a) with the highest content of quartz and feldspar are characterized by maximum β-values. Compressibility is also dependent on the size of grains. Coarse-grained rocks have a higher compressibility. Even in hydrostatic conditions these rocks are apt to a certain disintegration that increases the compression modulus at the further growth of pressure. The presence of plagioclase does not contribute to the increase of compressibility, for this mineral is rather plastic, the plasticity growing with pressure. Under hydrostatic compression, the grains of plagioclase tend to change their shapes within the same volume. Therefore, their effect on the β-values is not significant. This is indicated by compressibility of intermediate and acid rocks.

In the high-pressure experiments, investigation of the Poisson ratio (σ) behaviour is of a special interest. It is known that, in natural conditions, σ is higher in plastic bodies than in brittle ones. Therefore, the Poisson's ratio is helpful for a qualitative appraisal of plasticity and brittleness of rocks under the high pressure effect. Since σ characterizes the solid's ability for orthogonal deformations it can as well be indicative of redistribution of the internal stresses within the sample volume under high pressures [98].

The above considerations may serve a ground for the analysis of the state of rocks of different composition under hydrostatic compression. The rocks of acid composition are characterized by a wide range of the Poisson's ratio values (from 0.10 to 0.35) (fig. 7a). This is due, in the first place, to considerable variation of the mineral composition, porosity and texture-structural features of the rocks. Minimum σ-values are characteristic of the most brittle varieties rich in quartz. These are quartzite, equigranular granite and charnokite. Besides, low σ-values are found in the samples of coarse-grained rapaktivi granite and Chudnovo-Berdichevsky granite. The rocks with a higher content of plagioclase and potash feldspar have high σ-values. Under high pressures, the Poisson's ratio of acid rocks behaves variously. However, certain regularities can be established. So, the rocks with a high content of quartz (quartzite, for example) have a slightly changing σ (fig. 7a, s. *1*) and their plasticity increases with pressure insignificantly. The medium-grained rocks with a moderate content of quartz are characterized by extremum areas of the σ change. The higher is the quartz content the more prominent are the peak values of σ (fig. 7a, s. *4*, *6*, *8*, *10*). This may be explained like follows. It has already been discussed [48, 106—108] that under hydrostatic pressure the rock sample has a strain-stressed state due to inhomogeneous deformation of the rock-forming minerals. With increasing content of quartz — the most brittle, rigid and sufficiently isotropic mineral — the inhomogeneity in the stress distribution grows. Here, the quartz grains serve as indicators trying to penetrate the surrounding less elastic (feldspar) and more plastic (plagioclase) minerals or destroy some of them (pyroxene, biotite) along layers. This brings along the loss of proportion between axial and lateral deformations and ($\sigma = E/2G$). In such conditions shear deformations grow, hence, the Poisson's ratio values decrease, thus reducing plasticity of the rock. The further increase of pressure, plasticity may grow again, the structural disturbances being temporary "cured" and the Poisson coefficient, naturally, increases (fig. 7a, s. 2). The σ increase with pressure is favored by the enrichment of the rock with plastic minerals (plagioclase, pyroxene), the $\sigma = f(P)$ dependence being expressed by smooth curves similar to the V_p and V_s profiles (fig. 7a, s. *3*, *8*, *11*). In intermediate rocks, (fig. 7b) σ_{001}-values at atmospheric pressure and $\sigma = f(P)$ changes are similar to the data for acid rocks. In the rocks with a higher content of plagioclase the σ-values grow within the whole range of pressure (fig. 7b, s. *1*, *5*, *7*). Increase of the quartz and anisotropic biotite content in the rock causes a decrease of the Poisson's ratio values with pressure (fig. 7b, s. *2*, *6*).

While acid and intermediate rocks show mostly about the same σ-values under atmospheric and high pressures ($\sigma_{001} \approx 0.22$—$0.26$ and $\sigma_5 \approx 0.22$—0.32), the Poisson's ratio in basic rocks has somewhat higher values ($\sigma_{001} \approx 0.25$—$0.30$; $\sigma_5 \approx 0.27$—0.35) (fig. 7c). σ-values of basic rocks gradually grow with pressure. Plots of $\sigma = f(P)$ are essentially devoid of any anomalous sections. It is explained, in the first place, by structural features and mineral composition of the rocks. All the rocks under study are fine-grained with a higher content of plastic minerals (plagioclase, pyroxene). The behaviour of the $\sigma = f(P)$ plots for basic rocks suggests that under growing pressure these rocks tend to a more plastic state compared with acid and intermediate rocks.

Comparison of the experimental results obtained under quasihydrostatic and hydrostatic pressures and room temperature

It is known that the pressure distribution in the Earth's interior is not necessarely hydrostatic. At great depths, deviation from the hydrostatic regime may reach 10 and more percent. This, undoubtedly, affects the peculiarities of the elastic parameters change with depth. The results derived under hydrostatic pressure and those obtained under quasi-hydrostatic pressure have been compared by many investigators. However, this was usually done without consideration of inhomogeneities in the pressure distribution within quasi-hydrostatic chambers. The values may reach 10—15% and change, undoubtedly, the elastic characteristics of the sample under study compared with those found in hydrostatic conditions.

In view of the aforesaid situation a series of experiments were performed aiming at examination of the effect of irregularities in the stress distribution on the elastic parameters of rocks under high pressure. The samples under study were identical to those previously investigated under hydrostatic pressure to 5 kbar. In fig. 8—10 the most typical results of $V_{p,s} = f(P)$ and $\sigma = f(P)$ measurements under hydrostatic and quasi-hydrostatic conditions are compared and the errors of the parameters determinations shown. The velocity against hydrostatic and quasi-hydrostatic pressures plots behave differently and show certain regularities. The difference has appeared to be the most striking in coarse-crystalline rocks. In this case, velocities change smoothly with pressure (fig. 8, 9), the velocity values being somewhat higher under the hydrostatic loading than under quasi-hydrostatic. This is due to the increase of elastic deformations developing in the sample under additional axial loading characteristic of quasi-uniform pressure chambers of the "cylinder-piston"-type, especially in the initial pressure range ($P < 2$), and causing

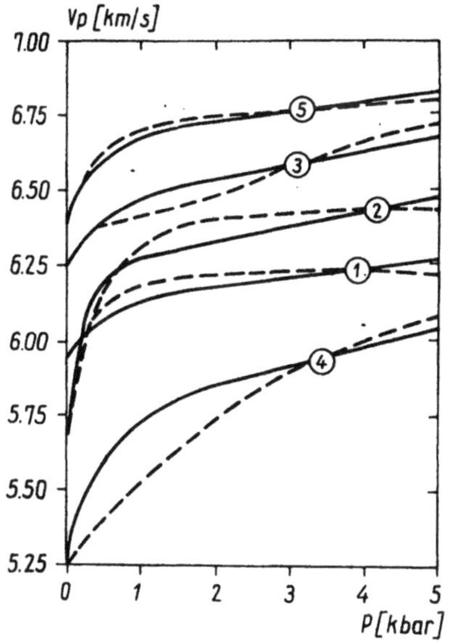

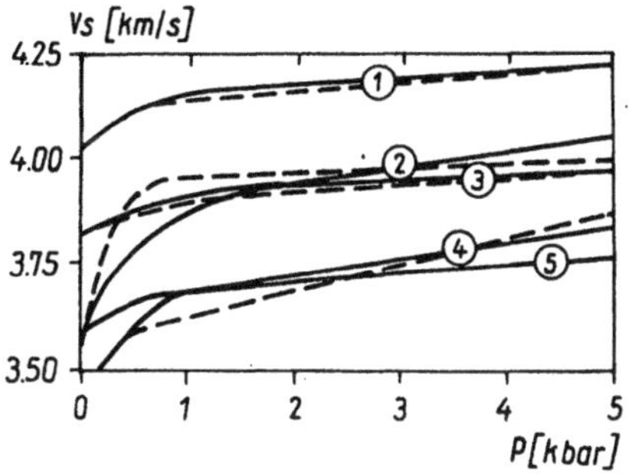

Fig. 9. Data von V_s change under hydrostatic (solid lines) and quasi-hydrostatic (dashed lines) pressures in different rocks. Same legend as in fig. 8.

Fig. 8. Data on V_p change under hydrostatic (solid lines) and quasi-hydrostatic (dashed lines) pressure in different rocks.
1 — quartzite; *2* — porphyry granite I; *3* — charnokite II; *4* — rapakivi granite II; *5* — labradorite II.

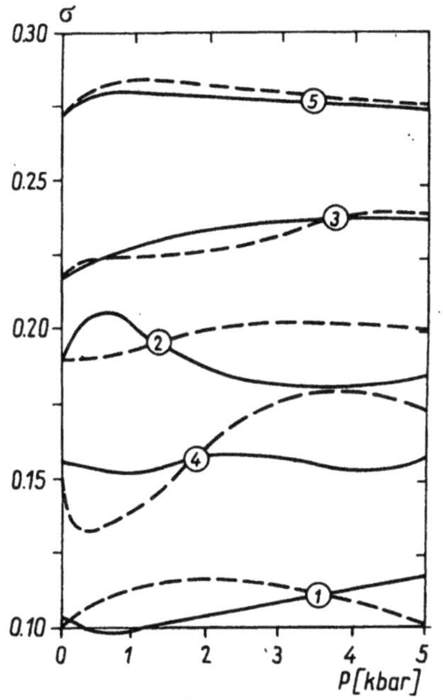

Fig. 10. Data on Poisson's ratio change under hydrostatic (solid lines) and quasi-hydrostatic (dashed lines) pressures in different rocks. Same legend as in fig. 8.

disarrangement of the original structure of the rock. A similar picture is observed in medium-grained varieties rich in quartz and with porosity more than 1%. Here, however, the difference is not so conspicuous (fig. 8, 9, s. *3*).

Let us, further, consider the character of the Poisson coefficient against pressure plot (fig. 10, s. *3*, *4*). In coarse-grained and porous rocks, in the initial range of pressure, σ is dropping sharply. Axial stress along the generatrix of the sample grows and brittleness increases. On reaching the limit value the stresses relax to cause different structural changes in the sample (brittle-plastic micro-disarrangements of the rock structure). With further increase of loading the predominant axial pressure drops and the sample acquires a state of hydrostatic compression. But since the structural disarrangements favour the subsequent compaction of the rock its elastic parameters become higher than in similar rocks under hydrostatic pressure.

Fine- and medium-grained rocks of low porosity show a considerably smaller difference in velocities measured under hydrostatic and quasi-hydrostatic pressure; the velocity values being somewhat higher in quasi-hydrostatic conditions Fig. 8, 9, s. *1*, *2*. The phenomenon can by explained by the data on the Poisson's ratio change (fig. 10, s. *1*, *2*). Plastic deformations originating in these rocks reduce the effect of brittle disarrangement of the rock due to irregular stress distribution in the working volume of the quasi-hydrostatic chamber (P_q). In the initial P_q range σ grows reflecting thus an increase of plasticity of mineral matter, and intensify compaction of the latter. The elastic parameters values grow more effectively than under hydrostatic pressure.

Coarse-grained monomineral formations and single minerals and fine-grained rocks behave similarly (fig. 8, 9, s. *1*). In the initial pressure range (to 3 kbar), the additional axial stresses in the quasi-hydrostatic pressure chamber favour a sharper increase of elastic constants but as soon as certain *P*-values have been achieved the elastic characteristics start decreasing (fig. 10, s. *5*) at the expense of the destructive effect of the internal anisotropic stresses within the volumes of single crystals.

The additional stresses in the rock samples resulted from deviation from hydrostatic conditions of the experiment can, thus, cause an increase or decrease of the elastic parameters of rocks. These phenomena are related to texture-structural features, porosity and mineral composition as well as to duration of the pressure application in the course of the experiment. Macrocrystalline medium-grained acid rocks rich in quartz, show a decrease of elastic parameters in the initial quasi-hydrostatic pressure range compared to the hydrostatic regime. This is due to the additional stresses developing in quasi-hydrostatic conditions. With the further increase of pressure and decrease of irregularities in the stress distribution the values of elastic constants are nearing to those observed in hydrostatic conditions. Fine-grained intermediate and basic rocks under additional stresses usually improve their elastic characteristics. Up to 3 kbar, the difference between the elastic parameters values measured in alike rocks of different composition under hydrostatic and quasi-hydrostatic pressures in usually decreasing and, mostly, does not exceed the error of determination. However, with the growth of quasi-hydrostatic pressure and decrease of inhomogeneity in the stress distribution this effect disappears. Beyond 3 kbar, this difference, measured in the hydrostatic and quasi-hydrostatic devices decreases and does not, usually, exceed the error of determination.

The influence of high temperatures under atmospheric pressure

The change of elastic parameters of mineral matter in temperature fields at different depths can be estimated only by a through examination of the temperature effect under varying pressure starting from the atmospheric [6, 7, 89, 116, 124, 126, etc.]. The influence of high temperatures on V_p and V_s variations in the atmospheric conditions were studied in the samples similar to those examined under high hydrostatic pressure.

Results of the experimental studies of $V_p = f(T)$ are presented in fig. 11a, b, c. Average values of the velocity decrease at 20 to 400 °C in acid rocks ($\Delta V_p/V_p$ and $\Delta V_s/V_s$) are, respectively, 45 and 40%; in intermediate rocks —40 and 35 and in basic rocks —30 and 25%. In quartzite (fig. 11a, s. *1*), in the same temperature range, the velocity change is only 10% as it is in glasses and fused quartz.

The differences in $V_{p,s} = f(T)$ behaviour for the three groups of rocks have proved to be not only quantitative. In acid rocks to 100 °C velocities fall with small gradients (in the order of 0.002 for V_p and 0.001 km/sec · °C — for V_s). Further, to 300 °C, an area of the largest velocity drop can be distinguished ($dV_p/dT \approx 0.010$, $dV_s/dT \approx 0.005$ km/sec · °C). Beyond 300 °C velocities decrease more smoothly ($dV_p/dT \approx 0.006$; $dV_s/dT \approx 0.002$ km/sec · °C). For intermediate rocks only two areas of V_p and V_s drops are distinguished. The first is in the range from 20 to 250 °C ($dV_p/dT \approx 0.008$; $dV_s/dT \approx 0.006$ km/sec · °C), the second is from 250 to 400 °C ($dV_p/dT \approx 0.006$; $dV_s/dT \approx 0.005$ km/sec · °C).

In basic rocks velocities change in the same way as in acid rocks, but the first zone of a gradual decrease of V_p and V_s is sometimes bounded by 200 °C ($dV_p/dT \approx 0.003$;

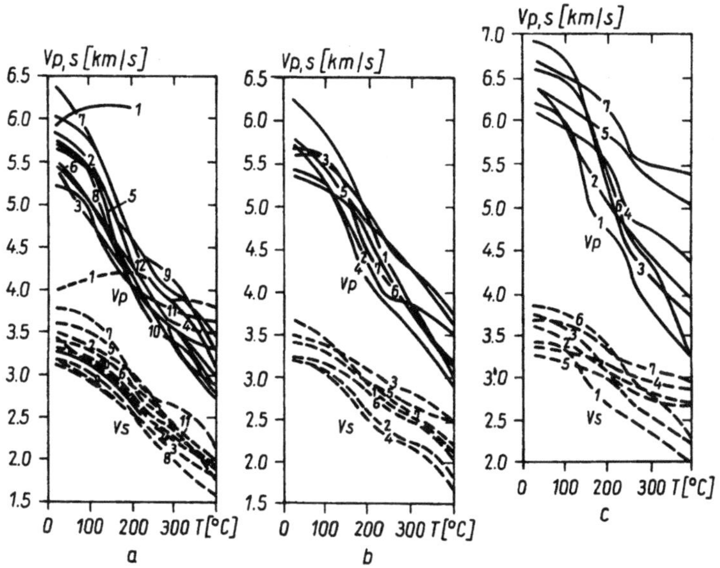

Fig. 11. V_p and V_s change with temperature under atmospheric pressure.
a — in acid rocks: *1* — quartzite; *2, 7* — charnokites I, II; *3, 6* — Zhitomirsky granites I, II; *4* — Kirovogradsky granite; *5, 9* — porphyry granites I, II; *8, 10* — rapakivi granites I, II; *11* — Chudnovo-Berdichevsky granite; *12* — Klesovsky granite. b — in intermediate rocks: *1* — charnokite III; *2* — sobite; *3* — granodiorite; *4* — Osnitsky granite; *5* — Klesovsky porphyrite; *6, 7* — diorites II, III. c — in basic rocks: *1, 2* -- labradorites I, II; *3* — gabbro norite; *4* — diabase; *5* — pyroxenite; *6, 7* — eclogite-like rocks I, II.

$dV_s/dT \approx 0.002$ km/sec · °C). Considerable gradients of the velocity change are found within an interval of 170—250 °C; $dV_p/dT \approx 0.020$ and $dV_s/dT \approx 0.008$ km/sec · °C. Later, they drop to $0.003 \div 0.010$ and $0.002 \div 0.005$ km/sec · °C, respectively.

The established regularities of $V_{p,s} = f(T)$ can be conveniently discussed using the data on the computed elastic characteristics shown in fig. 12—15. It is known that the main factor causing decrease of elastic parameters under heating is a disarrangement of the natural structure of the rock at the expense of an irregular and anisotropic expansion of

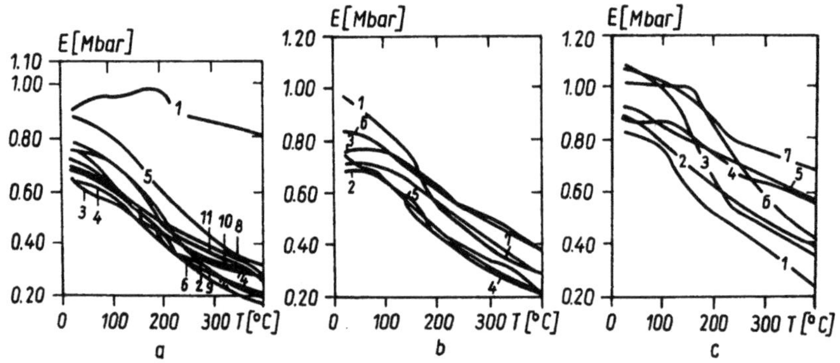

Fig. 12. The Young modulus against temperature plot under atmospheric pressure.
a — acid rocks: *1* — quartzite; *2* — charnokite I; *3, 6* — Zhitomirsky granites I, II; *4* — Kirovogradsky granite; *5, 8* — porphyry granites I, II; *7, 9* — rapakivi granites I, II; *10* — Chudnovo-Berdichevsky granite; *11* — Klesovsky granite. b — intermediate rocks: *1* — charnokite III; *2* — sobite; *3* — granodiorite; *4, 7* — diorites I, III; *5* — Osnitsky granite; *6* — Klesovsky porphyrite. c — basic rocks: same legend as in fig. 11c.

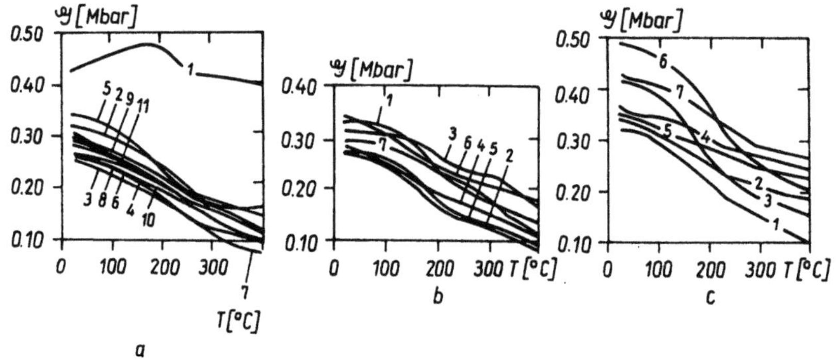

Fig. 13. The shear modulus against temperature plot under atmospheric pressure.
a — acid rocks; b — intermediate rocks; c — basic rocks. Same legend as in fig. 12a, b; 11c.

constituent minerals in temperature fields. In this connection the thermal expansion coefficient data for some minerals shown in fig. 16 may be of use [42, 95]. As is seen, the highest values of $\Delta\alpha/\alpha$ are inherent in quartz, the lowest in plagioclase with microcline holding an intermediate position. Hence, the rocks rich in quartz are characterized by maximum change of elastic parameters and the rocks with a high plagioclase content by minimum. Multicomponent anisotropic formations would, evidently, show a significant change of

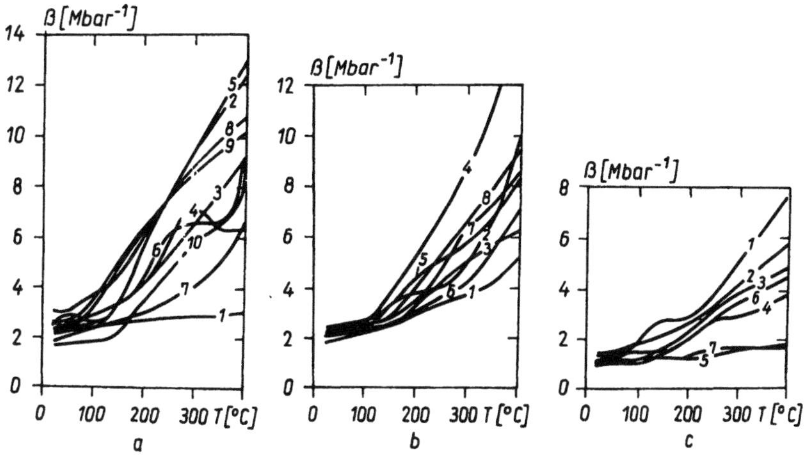

Fig. 14. Compressibility against temperature plot under atmospheric pressure.
a — acid rocks: *1* — quartzite; *2* — charnokite I; *3, 5* — Zhitomirsky granites I, II; *4* — Kirovogradsky granite; *6, 8* — rapakivi granites I, II; *7* — porphyry granite II; *9* — Chudnovo-Berdichevsky granite; *10* — Klesovsky granite. b — intermediate rocks: *1* — charnokite III; *2* — sobite; *3* — granodiorite; *4, 7, 8* — diorites I, II, III; *5* — Osnitsky granite; *6* — Klesovsky porphyrite. c — basic rocks: same legend as in fig. 11 c.

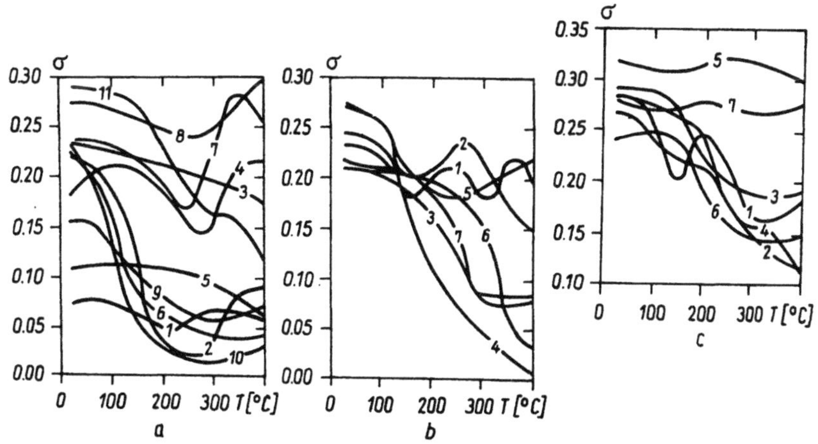

Fig. 15. Poisson's ratio — temperature variation under atmospheric pressure.
a — acid rocks; b — intermediate; c — basic rocks. Same legend as in fig. 12a, b; 11c.

elastic parameters. This is confirmed by the experimental results. For example, acid rocks with a large amount of quartz show the largest change of ultrasonic velocity with temperature. Basic formations composed mainly of plagioclase are less effected by temperature. But the same rocks enriched considerably by quartz and garnet (fig. 11c, s. *7*) are characterized by significant velocity change with temperature.

In spite of the different character of V_p and V_s decrease with temperature (from 50 to 400 °C) shown by deposits of different composition, the changes of the Young modulus (E) and coefficient of shear (G) remain, as for as the experiments show, almost the same for all the rocks (0.13 and 0.06 Mbar/°C, respectively).

The state of the rock under heating is more difinitely characterized by a change of compressibility and the Poisson's ratio. It is common knowledge that compressibility decreases with basicity and increases with the quartz content. At high temperature the change of compressibility depends also on mineral composition, texture-structural features and porosity. On the whole, compressibility of all the rocks increases with temperature, the biggest β change being found in acid (fig. 14a), the smallest — in basic rocks (fig. 14c). For the rocks of the same composition the compressibility increase is the more considerable the higher is porosity and content of secondary minerals. This suggests that, under the effect of temperature, the rocks with low values of elastic characteristics are more willing to disarrange their structure and become more friable than the rocks that have not undergone a secondary transformation.

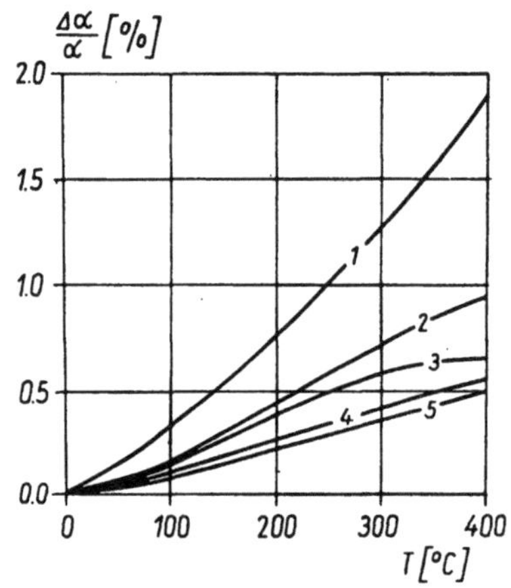

Fig. 16. Relative change of mineral dilation with temperature.
1 — quartz, *2* — granite, *3* — microcline, *4* — plagioclase (No 44), *5* — plagioclase (No 23).

This phenomenon is quite natural since the changed rocks have an imperfect elastic contacts between constituent minerals. With growing temperature the elastic properties of the rock in contact zones degenerate more intensively. Therefore, it may be suggested that in active tectonic zones the temperature influence on the change of elastic parameters is more significant than in stable areas. Another factor affecting the elastic properties behaviour under varying temperature is the size of grains. So, it is established that the larger grains the more essential is the change of compressibility. In the monomineralic rocks the compressibility grow, with temperature is less conspicuously than in multicomponent formations (fig. 14a, s. *1*).

To evaluate the mechanism of the change of elastic characteristics of the rocks under high temperatures due to redistribution of the internal stresses within the sample let us consider the interrelation between the Poisson's ratio and temperature (fig. 15). The change of this parameter is governed by a sofisticated law determining the behaviour of thermostresses under heating. Acid rocks have a wide spectrum of σ values within the temperature range from 20 to 400 °C (fig. 15a). Intermediate rocks are characterized by a narrow bound of σ_0-values ($0.18 \div 0.27$) at the room temperature which widens considerably to 400 °C ($\sigma_T = 0.02 \div 0.22$) (fig. 15b). On heating basic rocks the σ_0-values almost do not scatter (fig. 15c; $\Delta\sigma_0 = 0.08$, $\Delta\sigma_T = 0.09$). The behaviour of $\sigma = f(T)$ varies not only in rocks of different composition but also in rocks of the same name. The latter is due to the peculiarity of the rock structure and diversity of mineral compositon. Certain sections of the $\sigma = f(T)$ plots show the regularities characteristic of almost all the rocks under study. From 20 to 100 °C, σ usually grows slightly and gradually. In some cases the increases is due to thermoelastic compaction of the rock (fig. 15a, s. *4*; fig. 15b, s. *7*; fig. 15c, s. *3*, *6*). In other cases a decrease of the Poisson ratio is observed caused by low-temperature relaxation of thermoeleastic stresses and, hence, weakening of the mineral matter. In this interval of temperature the $\sigma = f(T)$ behaviour reflects a certain thermostability of constituent minerals and the water content in the rock. The change of volume of adsorbed water and transformation of the latter into a new state bring the water into an active movement within the sample. The water penetrates gentle microfractures towards the sample surface or it localizes, till a certain temperature, within the sample volume to produce an expanding effect weakening of molecular bonds. The water redistributed in this manner, under temperatures less than 100 °C, may in some cases affect an increase of the rock plasticity (σ is growing). Localized within a closed volumen (at about 100 °C and above) it enhances the existing thermostresses. This results in overstrengthening of certain parts of the sample. The rock becomes more brittle (σ is decreasing).

The further heating to 250—280 °C is usually accompanied by the greatest change of elastic parameters (especially for acid an intermediate rocks). Volume expansion of rock-forming minerals leads to reduction of space between interfaces of certain minerals. Deformation forces of the latter increase at the expense of the growth of thermoelastic stresses. The binds are getting stronger in this case. But the growing deformations reduce plasticity of the rock and therefore the Poisson ratio decreases.

Under growing temperature anisotropic thermal expansion occurs in minerals with different coefficients of linear thermal expansion along different crystallographic lines (fig. 16). This calls forth a constant development of relaxation phenomena on grain boundaries in the form of viscous slipping of constituent minerals. The slipping causes deformation of some minerals and the entire sample. These phenomena increase with temperature irregularly. At a certain stage they become dominant and lead to brittle-plastic deformations of the grains of constituent minerals.

In some cases, under heating beyond 250 °C, the plasticizing effect of temperature may cause an increase of the Poisson's ratio (fig. 15a, s. *1*, *2*, *4*, *8*, *9*; fig. 15b, s. *1*, *2*, *5*; fig. 15c, s. *1*, *5*, *7*). Beyond 300 °C most likely is the secondary process of thermoelastic energy accumulation at the expense of enhancement of thermal oscillations of atoms in lattice of minerals. These affect the mechanical contact between minerals. Cooling the samples from 400 °C to 300—250 °C proves this suggestion. That is, with decreasing temperature, the changes of elastic parameters have appeared the same as under the temperature increase. Therefore, within the said interval of temperature the changes of elastic characteristics are mainly due to mineral composition rather than rock structure and its disarrangements, under heating. The disarrangements are known to be irreversible.

The above suggestion about the influence of temperature below 400 °C on the elastic state of the rock is supported by the data from a thorough examination of the sample sections before and after heating. In some minerals the occurence of microjointing is established (the most conspicuous in quartz and feldspar). In quartz, intensive wavy and sometimes blocklike fading is observed pointing the mechanical deformation of the mineral. It is noted that coarse (more than 0.3 mm wide) grains of quartz undergo more intensive changes. The structure of small grains remains almost unchanged. Besides, deformation and migration of grain boundaries are revealed, the shape of boundaries acquiring serrated and sinuous configuration.

Simultaneous effect of high temperature and quasi-hydrostatic pressure

Longitudinal and transversal velocities and elstic constants (E, G, β and σ) for some rocks of the Ukrainian Shield have been studied experimentally under three different types of thermodynamic conditions corresponding to specific regions (fig. 17). Examined were the same acid, intermediate and basic rocks as considered in previous sections (Table 1).

The results of velocity studies in thermodynamic conditions of the first type are presented in fig. 18.

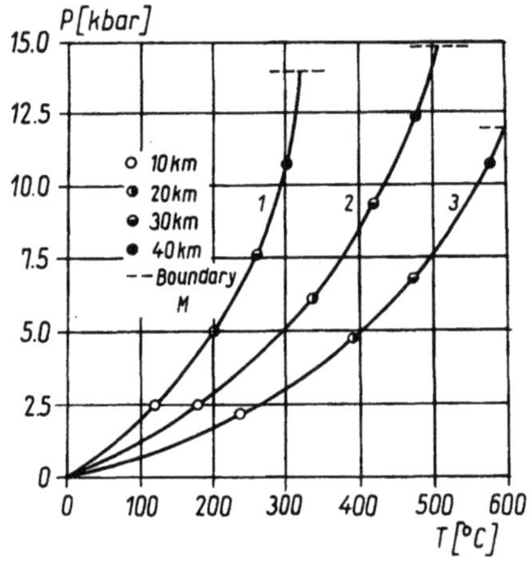

Fig. 17. Programs of experiments corresponding to pressure — depth and temperature — depth distributions for south-western (*1*), central (*2*), and south-eastern (*3*) parts of the Ukrainian Shield.

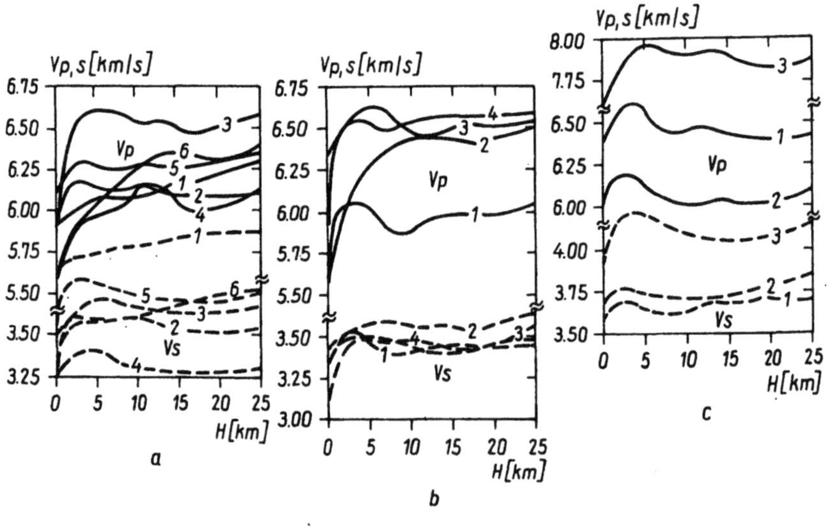

Fig. 18. V_p and V_s variation with depth.
a — in acid rocks: *1* — quartzite; *2* — charnokite I; *3* — Zhitomirsky granite I; *4* — Kirovogradsky granite; *5* — porphyry granite I; *6* — rapakivi granite II. b — in intermediate rocks: *1* — sobite; *2, 4* — diorites I, II; *3* — Osnitsky granite. c — in basic rocks: *1* — labradorite II; *2* — diabase; *3* — eclogite-like rock I.

Let us consider some general regularities in the velocity change with depth. Under the initial experimental PT ($H = 2 \div 5$ km) V_p and V_s grow as if temperature retains the room values. Upon reaching the maximum values they start dropping. The velocity decrease is generally observed to 7—10 km. Below this depth, in some cases, the velocity values grow. On the V_p, $V_s = f(P, T) = f(H)$ plots lower velocity zones are distinguished. They are most prominent in intermediate and basic rocks (fig. 18b, c). Further increase of PT-parameters (that is an increase of depth) has a varying effect on V_p and V_s change. On some velocity plots new extremum zones are observed. These are maxima corresponding to lower pressure (at depths of $12 \div 15$ km) and minima under higher pressure (at $15 \div 20$ km). On the plots without such secondary extrema, after the first maximum (H = 2—5 km), velocities decrease gradually and on reaching the minimum ($H = 14 \div 18$ km) start rising again. All the above regularities are less conspicuous for V_s.

Comparison between average values of V_p and V_s changes in acid, intermediate and basic rocks under various P, T and PT-regimes has revealed certain peculiarities. The greatest change of velocities under P = 5 kbar and T = 20 °C is found in intermediate rocks. These rocks are also characterized by a maximum change of V_p and V_s against temperature under atmospheric pressure. At the same time intermediate rocks show the most intensive growth of velocities under the simultaneous PT-effect. The applied pressure compensates almost completely mechanical changes of the said rocks possible under influence of high temperature. The difference between $V_p = f(P_5)$ and $V_p = f(P_5 T_{300}) = f(H)$ has proved, therefore, to be of a very small value (0.26 km/sec). Basic rocks show the least velocity pressure change compared with acid and intermediate rocks. At the same time velocity-temperature change in these rocks, is significant. In these rocks under simultaneous effect of P and T, the temperature influence is more effective than that of pressure: $V_p(P_{5\,\text{kbar}}) - V_p(H_{20\,\text{km}}) = 0.50$ km/sec. Therefore, V_p and V_s-values at depths of about 20 km are equal to those at the ground surface. In acid rocks, velocity changes under various thermodynamic conditions are in intermediate position regarding those in intermediate and basic rocks.

The observed regularities of V_p and V_s behaviour with P, T and PT are due, in the first place, to peculiarities of mineral composition and structure of the rocks. So, medium-grained varieties with the most widely varying mineral content and the highest enrichment in plagioclase (intermediate rocks) show the largest velocity change with PT parameters at the expense of the dominating effect of compensating pressure. Coarse-grained acid rocks rich in quartz and eclogite-like basic rocks are less sensitive to compensating external pressure. This, as mentioned, is due to irregular stress distribution more prominent in coarse-grained monomineral rocks under any kind of external loading. This suggestion is supported by the data on dynamics of the change of E, G, β and σ under a simultaneous P and T effect. Fig. 19—22show the plots of these parameters against depth for intermediate, acid and basic rocks obtained by the PT-programm of the first type (fig. 17).

Let us compare the Young modulus values computed from the V_p and V_s data under hydrostatic pressure and room temperature with the experimental results obtained at a simultaneous effect of quasihydrostatic pressure and high temperature. It is noted that with only P growing the modulus of longitudinal deformation grows smoothly in all the rocks varities under study, whereas under a simultaneous P and T effect the modulus values reach certain maxima at a depth of 3—4 km and, afterwards, either do not change at all or decrease to a minimum values at 12—18 km. In PT conditions corresponding to depths more than 20 km, the Young modulus mostly shows a new increase. The position of $E = f(H)$

minima depend on mineral composition of the rock. In acid rocks the minimum value corresponds to 17—18 km depth, in intermediate rocks — to 10—15 km, in basic — to 12—16 km. Under a simultaneous *P* and *T* effect the Young modulus values appear to be lower than under hydrostatic pressure. They are somewhat higher only in certain special cases (fig. 19a, s. *1*; fig. 19c, s. *3*). This is most clearly seen in eclogite-like rocks with the high quartz content, especially, when single crystals of quartz are in abundance. Thus, proportionality factor for the forces applied to magnitude of deformation along the main stress axis, corresponding to great depths, is actually dependent on the external compensation pressure. In this case, temperature surves as a plasticizing factor, although its change is sometime responsible for the existence of the low elasticity zones at depth from 5 to 20 km.

The regularities of the shear modulus (*G*) behaviour under various *PT* conditions

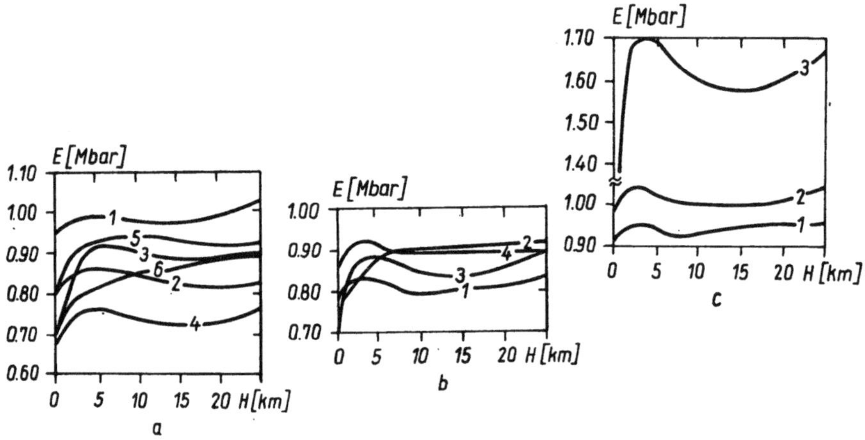

Fig. 19. Young modulus — depth variation.
a — in acid, b — intermediate and c — basic rocks. Same legend as in fig. 18a, b, c.

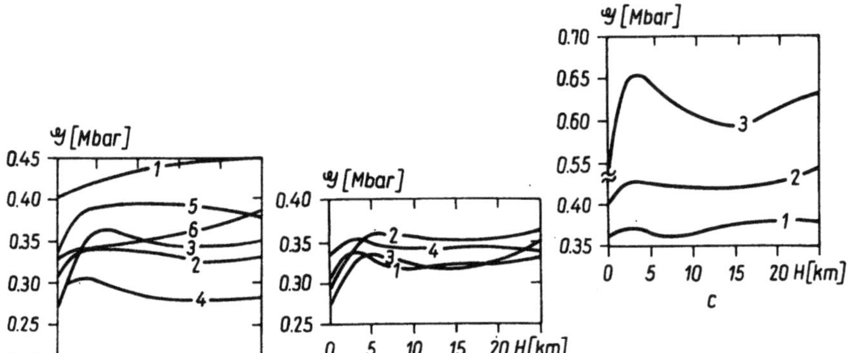

Fig. 20. The shear modulus — depth variation.
a — acid, b — intermediate, c — basic rocks. Same legend as in fig. 18a, b, c.

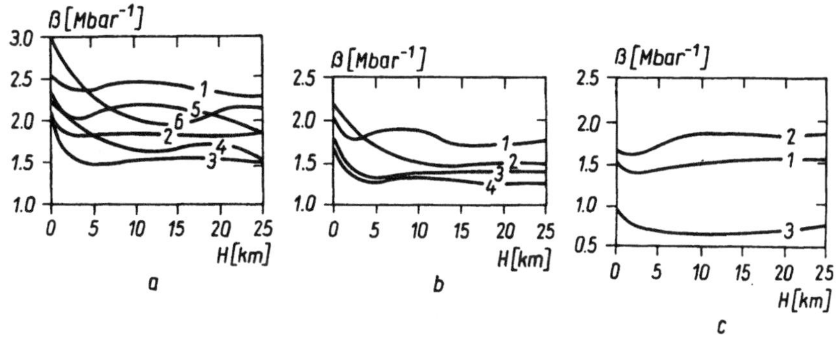

Fig. 21. Compressibility — depth variation.
a — acid, b — intermediate, c — basic rocks. Same legend as in fig. 18a, b, c.

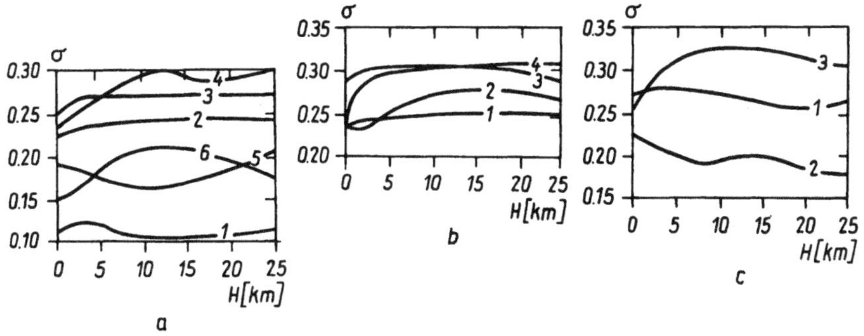

Fig. 22. Poisson's ratio — depth variation.
a — acid, b — intermediate, c — basic rocks. Same legend as in fig. 18a, b, c.

(fig. 20) are the same as for the Young modulus. However, they are pronounced less clearly. Hence, shear modulus (alike to V_s) is less sensitive to the effect of P and T.

Comparison of dynamic compressibility (β) data of rocks in varying thermodynamic conditions may be of interest (fig. 21). As it is known, compressibility is a characteristics of the ability of the rock to uniform compression and, hence, to the density change due to external uniform pressure. The high temperature (to 400 °C) experiments show 4—6 times increase of β resulted, evidently, from thermomechanical changes and increase of the rock plasticity. As to hydrostatic pressure, its 5 kbar value reduces compressibility no more than twice with density increasing to 2%. With temperature growth under atmospheric pressure as well as pressure increase under the room temperature, compressibility shows a gradual change without extremum values. Under a simultaneous PT effect compressibility changes by a sophisticated law sometimes, with reversal of the $d\beta/dH$ sign. On $\beta = f(H)$ plots, extremum areas are observed with anomalies of no more than 0.3 Mbar^{-1} at different depths. In terms of density this corresponds to its decrease by 0.02—0.05%. Softening of the entire rock caused by the temperature effect is but partly compensated by pressure. Under the simultaneous PT effect compressibility remains higher than under pure hydrostatic pressure. For example, for acid rocks under 5 kbar and 300 °C $\Delta\beta = \beta_H - \beta_P$ 0.7 Mbar^{-1} ($\Delta\varrho/\varrho \approx 0.1\%$ in terms of density). In intermediate and basic rocks $\Delta\beta$ drops to 0.3 and 0.5 Mbar^{-1}, respectively.

The behaviour of the rock compressiblity in the conditions of the simultaneous *PT*-effect points to rather complex processes operating under various dynamic regimes. The *PT*-conditions corresponding to depths of the order of 3—4 km cause consolidation of the rock. Sometimes temperature contributes as well. With *PT*-parameters corresponding to 5—10 km the *T* effect, in some cases, causes softening of the rock (compressibility grows). In this range of depths, the $\beta = f(H)$ plots show the zones of higher β-values (which are most prominent in acid rocks, fig. 21a). With further *PT* increase a second area of β decrease is observed. Hence, below 10—18 km the rocks start consolidating again.

To gain a deeper insight into the nature of the change of physico-mechanical properties of mineral matter at a great depth let us examine the regularities of the Poisson's ratio (σ) change under high pressure and temperatures (fig. 22). Below, compared are the results of $\sigma_p = f(P)$; $\sigma_T = f(T)$ and $\sigma_H = f(PT) = f(H)$-studies.

Polymineral rocks when heated to 250 °C loose their plasticity and become more brittle due to accumulation of anisotropic thermostresses within the samples. With further temperature

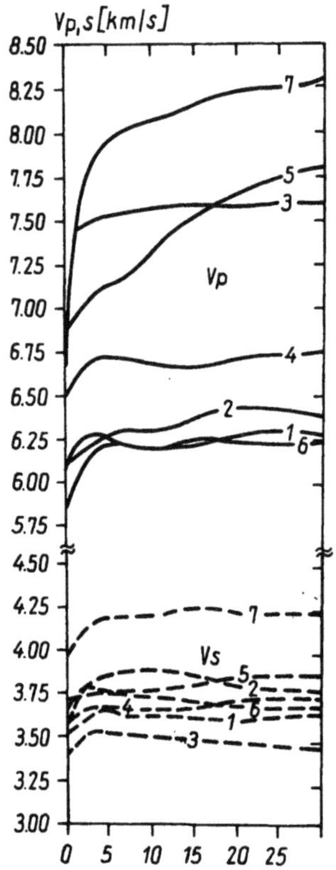

Fig. 23

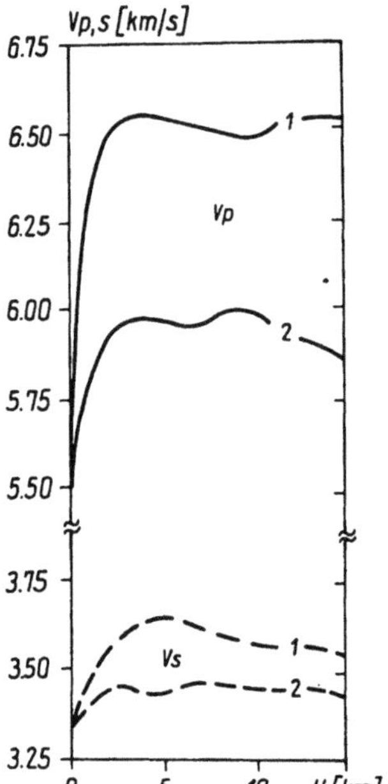

Fig. 24

Fig. 23. Longitudinal (V_p) and transversal (V_s) velocities variations in rocks of different composition under thermodynamic regimes of deep crustal zones under south-western part of the Ukrainian Shield.
1 — charnokite I; *2* — porphyry granite I; *3* — diorite II; *4* — labradorite II; *5* — gabbro norite; *6* — diabase; *7* — eclogite-like rock I.

Fig. 24. V_p, $V_s = f(PT) = f(H)$ plots for rocks of various composition in thermodynamic conditions of the crust under south-eastern part of the Ukrainian Shield.
1 — Zhitomirsky granite I; *2* — sobite.

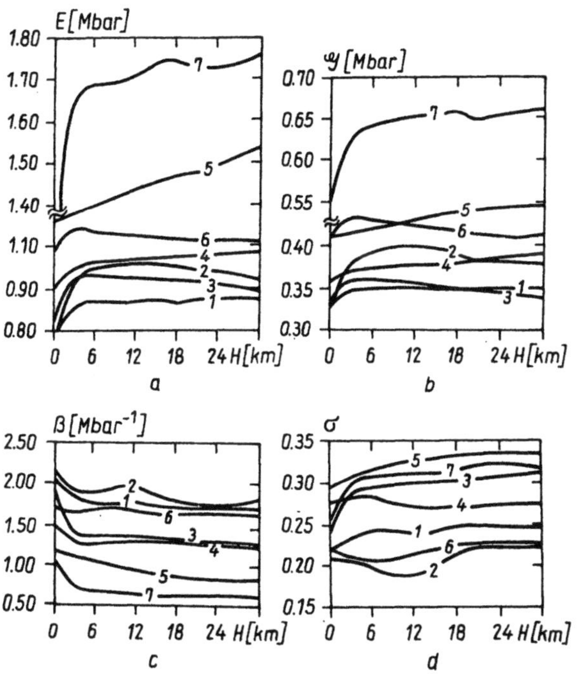

Fig. 25. Variations in Young modulus (a), shear modulus (b), compressibility (c) and Poisson's ratio (d) for rocks of different composition under thermodynamic regimes of deep crustal zones under south-western part of the Ukrainian Shield. Same legend as in fig. 23.

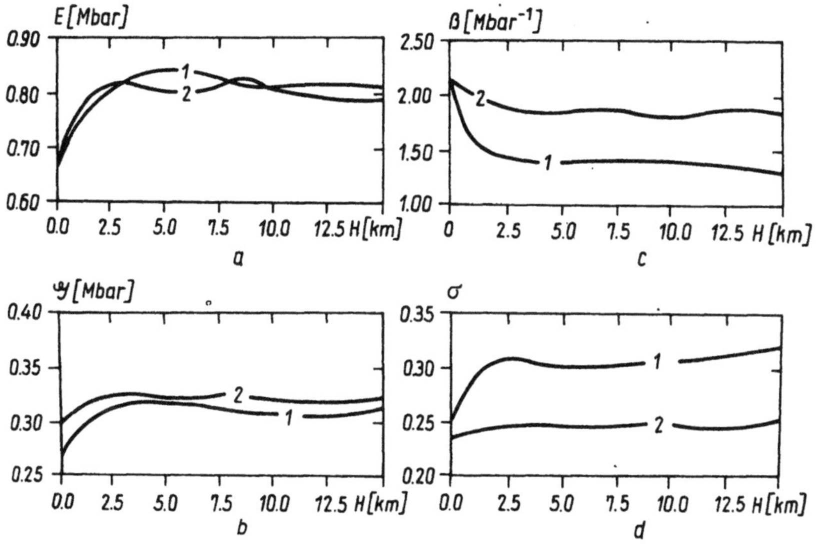

Fig. 26. Variations of the Young modulus (a), shear modulus (b), compressibility (c) and Poisson's ratio (d) in rocks of different composition under thermodynamic regimes of deep crustal zones under south-eastern part of the Ukrainian Shield. Same legend as in fig. 24.

increase the stresses relax into brittle plastic disturbances in the original structure of the rock. The latter become more plastic and, hence, more sensitive to external effects. Under room temperature an increase of pressure plasticizes the rock and, therefore, σ grows. On the basis of these facts one can qualitatively characterize the change that rocks undergo being as if submerged into various depth. For instance, in the initial *PT* range, corresponding to depths of 5—10 km (1 ÷ 2.5 kbar; 120—180 °C) the pressure effect dominates. The bonds between structural units within the rock improve and the rock becomes more dynamic and, hence, more plastic. Temperature just favours this process. So, even in less plastic rapakivi granite (σ_{001} = 0.15) the Poisson's ratio values are about 0.20 (fig. 22a, s. *6*) under the simultaneous effect of *P* and *T* corresponding to a depth of 7 km. Under hydrostatic and quasi-hydrostatic pressure (2 kbar) and temperature of 20 °C σ_2 equals 0.16. To 150 °C under atmospheric pressure the Poisson's ratio value drops to 0.10.

A plasticity increase with depth is characteristic of formations of any composition. However, at a depth of 10 km and deeper, rocks behave differently. Plasticity of certain rocks decreases with depth (σ_H decreases with *PT*). In this case, a simultaneous pressure and temperature effect does not produce a necessary relaxation of inhomogeneous stresses originating in the sample. The rock tends to a more elastic and, at the same time, more brittle state. This phenomenon is characteristic of coarse-grained and monomineral rocks, such as porphyry granite, rapakivi granite, labradorite (fig. 22a, s. *4*, *6*; fig. 22c, s. *3*). Equigranular rocks rich in plastic minerals (plagioclase and pyroxene) are less responsive to the change in thermodynamic conditions of the interior. This is confirmed by $\sigma = f(H)$ data for intermediate rocks.

Elastic characteristics of some rock samples were studied by a all the three experimental programs providing a simultaneous change of pressures and temperatures in accordance with the *PT* distribution within the crust under different parts of the Ukrainian Shield (fig. 17) [12, 15, 119, 122]. The high quasi-hydrostatic pressure and temperature apparatus [6, 7, 9] was designed at the Department of Physical Properties of the Earth's Material, at Institute of Geophysics, Academy of Science, Ukrainian SSR, Kiev. This equipment makes possible a modelling of thermodynamic conditions corresponding to depths to 30 km ($P \approx 7$ kbar, $T \approx 350$ °C). The results of the experiments are shown in fig. 23—26. Omitting the details of the behaviour of this or that parameter with depth under given *PT* regimes, let us compare the experimental results obtained for the same rocks by all the three programs. First of all it is noted that the rocks of different composition respond in a different manner to the change in experimental regimes. Acid rocks have appeared to be the most sensitive to variations of the *PT*-parameters.

The highest gradients of velocity increase with depth to 2—3 km are observed in the experiments conducted by the high-temperature program, the lowest recorded by the program of a low-temperature regime. Maximum velocity zones, if any, are found in the initial parts of $V_{p,s} = f(H)$ plots in connection with the given experimental regime. $V_{p,s}$ max are higher for low-temperature programs but they correspond to greater depths. In experiments conducted by high-temperature programs $V_{p,s} = f(H)$ plots are more complicated. Additional maxima and minima appear on them. The highest velocity deviations from mean values are found in experiments with high temperatures. In experiments with low temperatures, $V_{p,s} = f(H)$ are found to approach the same regularities observed under high hydrostatic pressures and room temperature. In this case, the higher dT/dH ratio and more complicated the rock composition the greater is the difference between $V_{p,s} = f(H)$ and $V_{p,s} = f(P_h)$. Similar dependence on the experimental conditions is established for E, G, β and σ (fig. 25, 26). As an example let us compare the data on the Poisson's ratio change obtained for the same rocks but in experiments conducted by different PT programs. In acid rocks, an increase of dT/dH entails decrease of brittleness of coarse-grained and monomineral formations under the same compensating pressure (fig. 22a, s. *2, 3, 5*; fig. 25d, s. *1, 2*; fig. 26d, s. *1*). As it has already been mentioned, intermediate rocks respond but slightly to the PT-parameters change. Consequently, the temperature effect on intermediate rocks is easily compensated by increase of pressure. This is shown by differently programmed experiments. So, in diorite and sobite, σ remains almost unchanged with depth, irrespective of P to T ratio (Table 1, s. *14, 19*), (fig. 22b, s. *1, 4*; fig. 25d, s. *3*; fig. 26d, s. *2*). In basic rocks plasticity grows with dT/dH as it does in acid rocks (fig. 22, fig. 25d, s. *4, 6, 7*).

Comparison of experimental results obtained under high hydrostatic and quasi-hydrostatic pressures and temperatures

Four samples from the Ukrainian Shield were subjected to examination on the hydrostatic pressure device under high temperatures [13]. These were porphyry granite re-

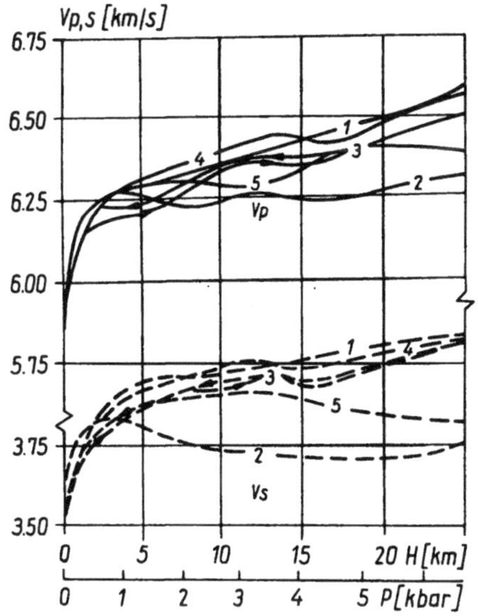

Fig. 27. V_p and V_s — variations in porphyry granite I under different PT — conditions of the experiment.

1 — V_p, $V_s = f(P_h)$; *2* — V_p, $V_s = f(P_q T) = f(H)$, thermodynamic regime of Earth's crust in central part of Ukrainian Shield; *3* — V_p, $V_s = f(P_h T) = (H)$, same th. r.; direct and reverse run of curve; *4* — V_p, $V_s = f(P_h T) = f(H)$, same th. r. but with min dT/dH; *5* — V_p, $V_s = f(P_q T) = f(H)$, th. r. of Earth's crust under south-western part of Ukrainian Shield.

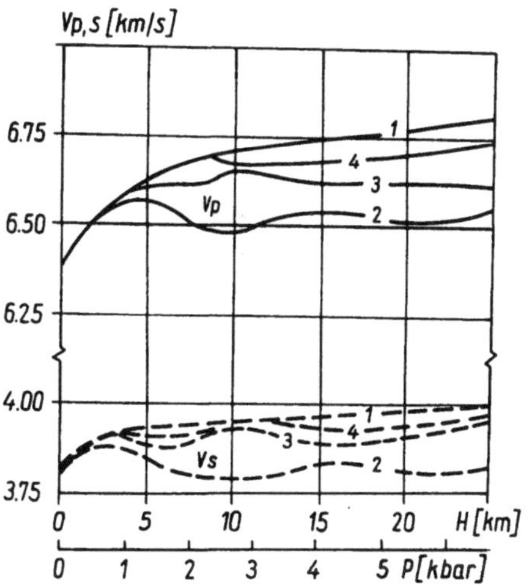

Fig. 28. V_p and V_s variations in charnokite II' under different PT — conditions of the experiment. Same legend as in fig. 27 (*1, 2, 3, 4*).

presenting acid rocks (fig. 27), Osnitsky granite and charnokite of intermediate (fig. 28, 29) and diabase of basic composition (fig. 30). The experiments were performed under two regimes provided by corresponding PT programs. One experiment studied the mean values of pressure and temperature change with depth characteristic of the central parts of the Ukrainian Shield ($H = 20$ km; $P = 5.3$ kbar; $T = 315$ °C). The other examined the same pressures but minimum temperature gradients ($H = 20$ km; $P = 5.3$ kbar;

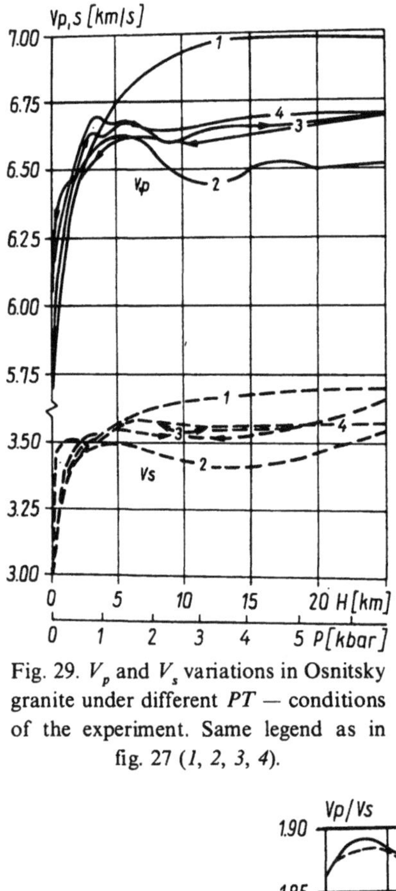

Fig. 29. V_p and V_s variations in Osnitsky granite under different PT — conditions of the experiment. Same legend as in fig. 27 (*1, 2, 3, 4*).

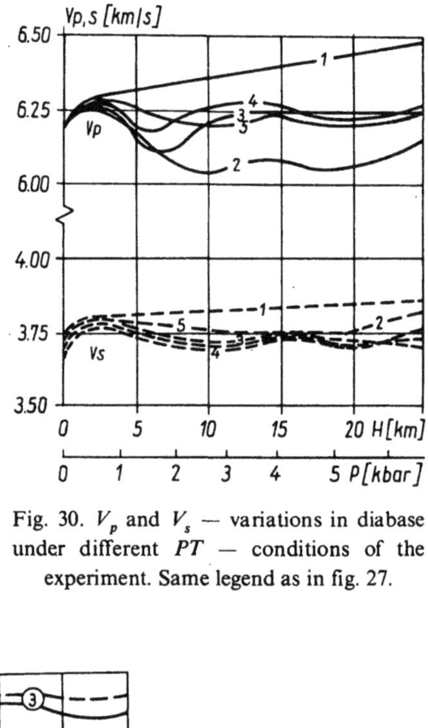

Fig. 30. V_p and V_s — variations in diabase under different PT — conditions of the experiment. Same legend as in fig. 27.

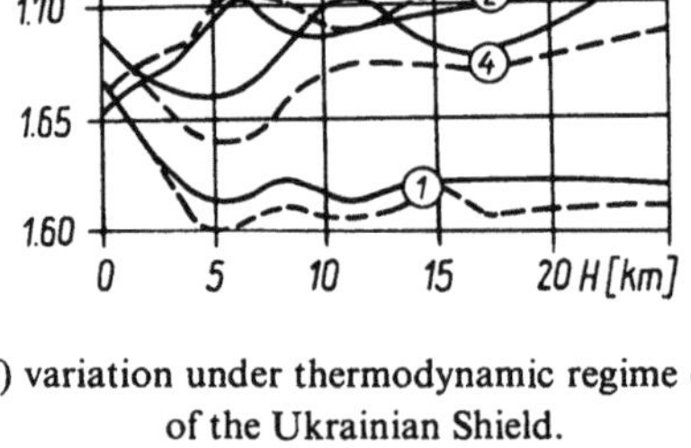

Fig. 31. $V_p/V_s = f(P_h T) = f(H)$ variation under thermodynamic regime of the Earth's crust in central part of the Ukrainian Shield.
1 — porphyry granite I, *2* — charnokite II, *3* — Osnitsky granite, *4* — diabase. Dotted curves mark thermodynamic regime with min dT/dH.

$T = 290$ °C). Longitudinal and transversal velocities were studied under PT regimes corresponding to depths to 25 km. Both the granite samples were studied under increasing and decreasing PT-parameters. The experimental results of V_p and V_s studies are shown in fig. 27—31, where $V_{p,s} = f(P_h)$ and $V_{p,s} = f(H) = f(P_q T)$ dependences obtained from the duplicate samples are presented for the sake of the convenience of the analysis.

Longitudinal and transversal velocities changes studied in the high pressure (both hydrostatic and quasihydrostatic) and temperature experiments is governed by a complex law (fig. 27—30). On the plots of $V_{p,s} = f(P_h T) = (H)$ several extrema are easily discerned. These maxima and minima are the most clearly expressed in diabase ($V_{p_{max}}$ is at 2 and 12 km; $V_{p_{min}}$ — at 16 and 18 km) and almost absent in charnokite and Osnitsky granite. It is worth mentioning that anomalous areas of the velocity change keep with the decreasing pressure and temperature as well, though they are somewhat smaller. This supports our suggestion that low and high velocity zones within the crust may be due to redistribution of thermoelastic and elastic stresses, caused by the effect of external pressures and temperatures within the volume of multicomponent polycrystalline aggregate.

In ideal PT-conditions that is under hydrostatic pressure and small dT/dH values the originating internal stresses cause reversible changes of elastic parameters. Consequently, under conditions of the interior, rocks are rather responsive to the change of thermodynamic regimes. Disturbance of a thermodynamic balance at a certain depth is not necessarily to cause a decrease of elasticity. This is proved by the reverse run of $V_{p,s} = f(H)$ plots (fig. 27, 29, curve *3*). Thus, the external hydrostatic pressure applied according to the current program compensate completely possible structure disturbances in high-temperature fields.

Besides, the sophisticated law of the $V_{p,s} = f(H)$ change is stipulated, to a considerable degree, by internal stress fields within the volume of the sample under study. These fields cause, sometimes, a complete reversibility of elastic parameters of the mineral matter. Under quasihydrostatic pressure (fig. 27, 30, curves *2*, *5*) the PT regimes of the experiments can cause a wide variation of $V_{p,s} = f(H)$ values. Similar phenomena are observed under hydrostatic pressures. In the experiments conducted by the same program but with somewhat lower (by 1.25 °C/km) thermal regime, higher velocities were recorded. In this case the greatest increase of V_p and V_s was observed within the depth range from 4 to 8 km to amount 125 m/sec for acid rocks, about 8 m/sec for intermediate and almost to 100 m/sec for basic rocks. It is once more emphasized that the intermediate rocks rich in plagioclase and moderate in quartz are less sensitive to variations of PT-regimes. Interesting results have been obtained for porphyry granite under low temperature experimental PT-regime. In this sample, V_p and V_s appear to be higher than under hydrostatic pressure and room temperature. This is, apparantly, due to the intensive consolidation of the multicomponent porous ($n = 1\%$) rock under temperature when the latter has not yet caused considerable thermoelastic stresses responsible for decrease of velocity.

In hydrostatic experiments, the high-temperature regime favors an increase of the rock plasticity (σ is growing), decrease of elastic modulus values (E and G) and growth of compressibility as it does under quasihydrostatic pressure (fig. 32). On comparing the absolute values of velocities (fig. 27—30) and other elastic characteristics of the rocks studied under high quasihydrostatic and hydrostatic pressures it is possible to estimate the influence of additional (nonhydrostatic) stresses on the change of physico-mechanical properties of rocks in the PT-conditions corresponding to great depths. Quasihydrostatic

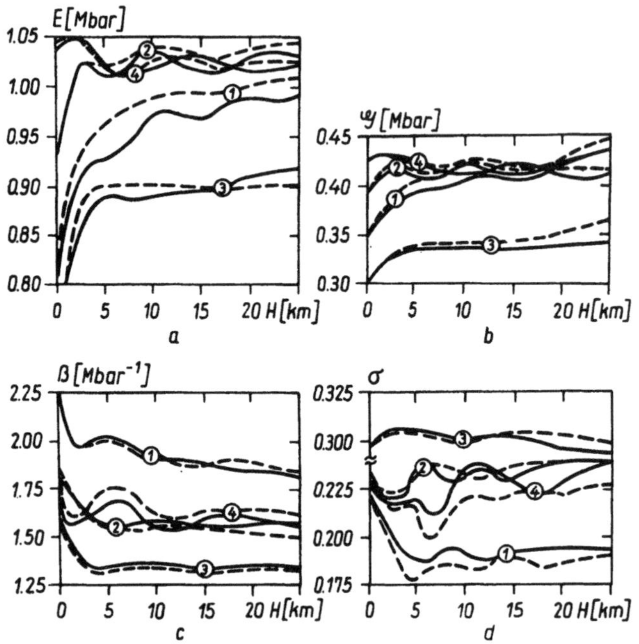

Fig. 32. Variations of the Young modulus (a), shear modulus (b), compressibility (c), Poisson's ratio (d) in rocks of different composition under thermodynamic regimes of the earth's crust in central part of the Ukrainian Shield (hydrostatic pressure).

1 — porphyry granite I, *2* — charnokite II, *3* — Osnitsky granite, *4* — diabase. Dotted lines designate thermodynamic regime with min dT/dH.

conditions with the same temperature regime cause a decrease of velocities, Young and shear moduli. In these conditions, compressibility grows and the Poisson ratio changes more smoothly. This is due to nonelastic microdeformations and disarrangement of the original structure of samples in the inhomogeneous stress fields.

Zones, sometimes extended, of extremum velocities change are distinguished more reliably on $V_{p,s} = f(P_q T)$ plots. Especially conspicuous is the first maximum which is not discernible in experiments with hydrostatic pressure. At the same time, some maxima on $V_{p,s} = f(P_h T)$ plots are almost invisible on $V_{p,s} = f(P_q T)$ curves. These maxima are resulted from increase of inner stresses originating within the rock due to high temperature and pressure. In quasihydrostatic conditions, these stresses relax more easily (under the effect of additional stresses) in the form of local brittle-plastic microde formations of minerals with the subsequent consolidation of the rock. Therefore, in quasihydrostatic conditions the velocity increase with PT-parameters is more rapid compared to that in hydrostic conditions up to the PT-parameters corresponding to depths around 20—25 km.

Some generalizations

The results of high pressure (quasihydrostatic and hydrostatic) and temperature experiments, including a simultaneous pressure and temperature effect, reveal a complicated distribution of physico-mechanical properties of rocks and minerals within the Earth's interior.

Hydrostatic pressure to 1—3 kbar (at room temperature) causes a considerable increase of elastic paramters at the expense of the variation in type, amount and volume of macrodefects (pores, fractures, intergranular boundaries, etc.). The closing of pores and fractures and improvement of intergranular contacts are caused by elastic and inelastic deformations of mineral matter. This is responsible for accumulation and redistribution of residual and induced microstresses within the sample volume, inhomogeneity of the microstresses increasing sometimes with pressure. The aforementioned phenomena favour activization of mobility of microdefects in the crystal (point defects, dislocations, localization of admixed atoms). In the most favourable conditions, a considerable amount of these defects abandon the major part of the mineral then improving elastic characteristics of the mineral association on the whole. With pressure growing, the changes of elastic characteristics are due, mainly, to atomic-molecular interactions of mineral components proper and to their elastic anisotropy. When constituent minerals are sufficiently anisotropic the further growing of pressure may cause an intensive accumulation of internal inhomogeneous microstresses which would relax in the form of brittle microdisturbances and plastic flows. The latter slowdown the increase of elastic parameters with hydrostatic pressure. It is worth noting that all these changes depend considerably on the original characteristics of the rock sample (its mineral composition, dimensions and type of porosity, the state of individual minerals, size and orientation of grains, etc.) and, naturally, the conditions of the experiment.

Our studies of some samples from the collection of the Ukrainian Shield rocks including acid, intermediate and basic varieties have proved the following regularity: longitudinal and transversal velocities, the Young and shear moduli grow with an increase of basicity. The relationship also holds under high hydrostatic pressures. Higher values of elastic parameters of intermediate rocks compared with those of acid rocks are due to the presence of pyroxene and hornblende within the rock. The high plagioclase content in intermediate rocks, along with insignificant amount of quartz, is responsible for a higher gradient of elastic parameters' increase with pressure compared to acid rocks. The high content of garnet and hornblende in basic rocks agrees with the highest elastic parameters of these rocks.

The high pressure effect on compressibility of rocks is also connected with composition. The highest compressibility is characteristic of acid rocks. The least tendency to the density increase with pressure in shown by basic rocks. The Poisson's ratio, characterizing the ability of solids for orthogonal deformations and reflecting qualitatively the brittle or elastic state of the body, makes possible a detailed analysis of dynamics of the elastic parameters' change in rocks of various composition under high pressure. More brittle are appeared to be the rocks with high quartz content. More plastic rocks are those characterized by a higher content of plagioclase and biotite. Therefore, acid rocks have lower values of σ and basic rocks have higher σ. Under high pressure, the acid formations tend to accumulate and subsequently relax the internal microstresses within the sample volume. This tendency is displayed in areas of maxima and minima on the $\sigma = f(P)$ plots. Basic rocks tend with pressure, to acquire a more plastic state. Therefore, the rocks rich in biotite, plagioclase, pyroxene and hornblende are more sensitive to the change of outside conditions, with no relaxation occuring.

The regime of the Earth's interior is not always characterized by a hydrostatic distribution of pressure. Nonhydrostatic character of the pressure effect on the rocks in the laboratory causes certain changes in elastic parameters due to texture-structural features of rocks, their porosity and mineral composition as well as the sample loading rate in the course of the experiment. Coarse-crystalline medium-grained acid rocks rich in quartz, under additional stresses in quasihydrostatic conditions, are characterized by lower elastic parameters under low pressures compared to hydrostatic conditions. As to fine-grained formations of intermediate and basic composition, their elastic parameters are improving in quasihydrostatic conditions. Beyond 3 kbar, the difference between elastic parameters measured in hydrostatic and quasihydrostatic conditions becomes insignificantly small. Hence, it is suggested that, in quasihydrostatic conditions, the upper crustal layers rich in quartz (at a depth to 10—15 km) are less plastic and tend to brittle deformations. As to intermediate and basic rocks composing deeper horizons, their plasticity increase under additional loading and the rocks become more dynamical and mobile.

Decrease of elastic characteristics under heating at atmospheric pressure is due mainly to disturbance of the natural structure of the rock resulted from irregular and anisotropic expansion of the constituent minerals in temperature fields. In this case, the rocks rich in quartz are characterized by maximum change of elastic parameters and the rocks with high plagioclase content — by minimum change. The greatest change is shown by multicomponent anisotropic formations. Elastic characteristics of rocks rich in secondary minerals and the rocks with imperfect bonds on the grain boundaries show an intensive change with temperature.

Variety in the changes of physico-mechanical properties observed in rocks under a separate effect of P and T is responsible for their rather complicated distribution under a simultaneous PT effect. In the uppermost layers of the crust (to 3—6 km) elastic parameters (V_p, V_s, E, G and K) of mineral matter increase monotonously with gradients dependent, mainly, on composition and texture-structural characteristics of rocks as well as on peculiarities of the pressure and temperature change with depth, the presence of inhomogeneous stresses within the crust and so on.

With the assumption that mineral formations at depths to 5—6 km have a constant composition, the highest gradients of V_p, V_s, E and G increase are characteristic of intermediate, the lowest — of basic rocks. In fine-grained basic rocks, the gradients of elastic parameters change with depth are higher than in coarse-grained rocks rich in quartz. In the crustal areas composed of equigranular acid and intermediate rocks with a high quartz content large gradients of elastic parameters increase are found with temperature gradient growing with depth. A similar effect is produced, on these rocks by additional stresses. The majority of intermediate and basic deposits within the depth range in consideration are indifferent to the change of temperature gradient (from 10 to 25 °C/km). However, the inhomogeneity of the pressure — depth distribution (less than 10%) causes a growth of elastic parameters with depth.

Compressibility of rocks in the upper crustal layers shows the most intensive change. High values of this parameter are found in acid, the lowest — in basic rocks. High density is inherent in rocks developed in areas with lower heat flow under hydrostatic distribution of pressure. The existence of additional stresses reduces the mineral matter density. This is most clearly manifested in acid formations. Intermediate rocks are less likely to soften.

An analysis of data on the change of the ratio of longitudinal to shear deformations (Poisson's ratio) yields a qualitative estimate of plasticity change with depth. To 5—6 km all the rocks irrespective of their composition tend to plastic state. According to our data, an exception is made by coarse-grained varieties rich in quartz. In the upper crustal layers, the mineral matter of intermediate and basic composition is believed to be rather plastic.

In areas with higher temperature gradients the rocks under quasihydrostatic pressure are characterized by a higher plasticity. In the case of hydrostatic distribution plasticity decreases with depth insignificantly due to predominance of thermoelastic stresses. It is, then, suggested that the change of elastic constants at the above mentioned depths is predominantly affected by pressure.

With mineral composition assumed constant the Earth's crust at the modelled depths of about 5—20 km is mostly characterized by the presence of low velocity zones. These have, probably, a thermodynamic nature and result from the opposite effects of pressure and temperature on the rocks. Since the latter have a complex structure and multicomponent composition, accumulation of elastic energy and subsequent relaxation may occur with a certain periodicity dependent on the mineral matter state at a certain depth, as well as the values of temperature and pressure gradients. Our experimental data have shown that several low velocity zones of different character can sometimes exist within the Earth's crust.

The thickness of low velocities zones vary they may be as thick as 10—20 km or thin (about 3—7 km). Thin low velocity zones are characterized by higher velocity gradients compared to thick ones. Thick low velocity zones are found mainly in basic and intermediate rocks, especially under quasihydrostatic pressure. In these zones, the velocity behaviour is strongly dependent on temperature gradient. In areas of low temperature gradient the velocity change under quasihydrostatic conditions is not sharp. In areas where temperature gradient is high, thick low velocity zones are also poorly recognizable against a background of a general velocity decrease. Under the circumstances the more clearly distinguishable are thin low velocity zones repeatedly occuring at shallow depth. These are more frequent in acid rocks and some intermediate rocks rich in quartz. It is noted that thin zones may be included into thick ones constituting as if a secondary structure.

Under hydrostatic distribution of pressure, thin low velocity zones are clearly distinguished at each 5—7 km of depth (the distance is measured between maximum values of $V_{p,s} = f(H)$). In this case velocity gradients are somewhat higher than at the same temperature but quasihydrostatic pressure.

The results of the experimental study of elastic parameters of rocks under various *PT*-conditions make possible to characterize the physico-mechanical state of mineral matter in low velocity zones. Here, lower values of the Young and shear moduli are characteristic of mineral matter, in the first place, as well as lower density (a higher compressibility). The data on the Poisson's ratio change suggest that the rocks enclosing low velocity zones are mostly characterized by more plastic state (compared to the central parts of the zones). The greatest change in V_p and V_s correspond to the maximum values of $\sigma = f(PT)$. The above suggestions apply to the crustal model of homogeneous composition. With the assumption of significant variation in mineral composition at depths of 5—30 km the regularities of the change of physico-mechanical state of rocks should be somewhat different. In acid rocks (of various composition and texture-structural pattern), low velocity zones are not distinguished. With basicity growing, the zones can be identified in mineral associations of widely varying composition. In intermediate rocks, low velocity zones are supposed to occur from 7 to 15 km deep ($\Delta V_p < 0.15$ km/sec in areas where $dT/dH \approx 15$ °C/km). Their thickness is less than 7—8 km. According to the $V_{p,s} = f(H)$ data intermediate rocks, may include low velocity zones thicker than 15 km at depth beneath $10 \div 20$ km ($\Delta V_p < 0.2$ km/sec). Low velocity zones of thermodynamic nature, then, probably, exist in thick masses of basic and intermediate material as well.

It is evident that at the boundaries of low velocity zones a gradient change of acoustical rigidity takes place. Under certain conditions, this can be detected by the methods of explosion seismology. Large low velocity zones have actually been stated in different geologic regions at depth of about 10—30 km within the continental crust.

General conclusions

1. On the example of the *PT*-programmed experimental study of mineral formations in the Ukrainian Shield it is shown that acid, intermediate and basic rocks under atmospheric and high pressures vary in the elastic parameters' behaviour. Acid rocks have relatively low values of V_p, V_s, E, G and high compressibility. Under high pressures, these rocks are less plastic and characterized by relaxational phenomena in the form of brittle-plastic micro-disarrangements of the structure of mineral matter. Under various pressure, basic rocks tend to gradual increase of elastic parameters and plastification without any anomalous phenomena.

2. Elastic characteristics of rocks behave differently under hydrostatic and quasihydrostatic pressures. The largest variations are characteristic of coarse-grained acid rocks, the smallest — of fine-grained basic rocks. The latter are rather plastic even in the conditions of inhomogeneous pressure distribution.

3. Under high temperature (atmospheric pressure) elastic parameters change, mainly, at the expense of the mineral matter softening caused by irregular and anisotropic expansion of constituent minerals and decrease of their elastic characteristics. The most willingly subjected to the change are coarse-grained formations of acid composition and certain rocks rich in secondary minerals.

4. The results of the experiments with a simultaneous effect of pressure and temperature suggest that in the upper horizons of the crust (to 3—6 km) the elastic parameters' values (V_p, V_s, E, G and K) can smoothly grow with depth due to incrase of the pressure effect. In these conditions, temperature affects mainly the growth of the mineral matter plastification.

5. The state of rocks under high experimental *PT*-regimes corresponding to depths from 5 to 30 km is characterized by decrease of the elastic parameter values compared with those found in high pressure experiments at room temperature. At this depth, the change of elastic properties is governed by complex laws with origination of low velocity zones. The character of the latter (position within the rock sequence, thickness and gradient value) are dependent on the pressure-depth distribution, temperature and mineral composition of rocks.

6. Thick (to 10—20 km thick) and thin (about 3—7 km) low velocity zones can be distinguished. Acid rocks are related to thin zones located mainly in the upper crust. Basic and intermediate rocks more often constitute thick zones. The areas with high temperature gradients include thin low velocity zones. Under condition of nonhydrostatic distribution of pressure, thick low velocity zones may mainly occur in the formations of basic composition.

7. The experimental data obtained suggest a thermodynamic nature of certain low velocity zones as derived from explosion seismology at depth of 10—30 km within the continental crust. These zones are composed of intermediate and basic rocks characterized by lower elasticity, lower density with higher brittleness in the marginal and higher plasticity in central parts of the zones.

CHAPTER II

Studies of Deformation Behaviour

INVESTIGATION OF PHYSICAL PROPERTIES OF ROCKS AT HIGH PRESSURE AND TEMPERATURE AS RELATED TO GEODYNAMIC PROCESSES

M. P. Volarovich

At present a great attention is given by geophysicists to the behaviour of the Earth's material under the pressure-temperature conditions which are believed to exist in the Earth's interior, and primarily in the crust and mantle. In this connection, in the last decades many countries launched into investigations of physical properties of rocks and minerals at high pressure and temperature. Thus, a new branch of the Earth's physics has come into being, known as high-pressure geophysics.

In 1950 the Laboratory of High Pressure Investigations at the Institute of the Earth's Physics of the USSR Academy of Sciences embarked on a research on physicomechanical and electrical properties of the mineral substance of the Earth. Various kinds of apparatus were created. The media chosen to transmit pressure were gas (nitrogen), liquid or plastic solids of lead, indium or pyrophyllite [57, 126]. The equipment is capable of measuring velocity and attenuation of elastic waves, elastic moduli and Poisson's ratio, compressibility, density, strength, plastic deformation, conductivity and dielectric constant in samples of rocks and minerals in various pressure intervals to several tens kbar. An improved experimental technique was developed allowing measurements at simultaneous high pressure and high temperature.

Elastic properties, density, strength and electrical properties of igneous, metamorphic and sedimentary rocks were studied at various pressure as a function of mineralogical composition, crystal structure, anisotropy, porosity (volumetric and slit-shaped), liquid content in pores, age and other factors. Concurrent mesurements of compressional velocity and density at pressures to 20 kbar, and statistical analysis of the numerous data obtained made in possible to establish a correlation between these parameters of rocks. The high correlation coefficient increased with pressure. It was found that compressional velocity decreased in the process of plastic deformation of rocks and minerals. The same phenomenon was observed in rock samples that had undergone in situ plastic deformations, such as dynamic metamorphism, serpentinization, cataclasis, etc. The effect of metamorphism on elastic properties as a function of pressure was studied for deep-seated rocks — periodotites, pyroxenites and eclogites. A few years ago, in cooperation with the scientists of the Geophysical Institute of the CzSSR Academy of Sciences (Prof. J. Wanek, Dr. Z. Pros, Lr. K. Klima) we made experiments on eclogites from Czechoslovakia under pressures to 4 kbar [128]. Good results have been recently obtained by the investigators of German Democratic Republic, Czechoslovakia, Poland, Hungary and the USSR in the research performed according to a general coordinated plan.

Elastic velocities and compressibility were studied in some rock-forming minerals under pressure to 15—20 kbar (in some experiments to 60 kbar). These experiments were run on quartz, feldspars, pyroxenes, amphiboles, calcite, nepheline, beryl and other minerals. The study of minerals belonging to various groups and classes showed that velocity variation with pressure was dependent on crystal directions. Velocities were found to increase more intensively in those directions which provide vacant possibilities for the distortion of a mineral's structure without breaking bonds, for instance, in the places where silicioxygen tetrahedron structures are separated by cavities.

The results of the research previously conducted at our Laboratory were reported and discussed in a number of paper, reviews and monographs [57, 58, 125, 127, 129, 130]. In these publications the reader will get aquainted with the history of investigations of physical properties of rocks and minerals under high pressures, and will find extensive references on the subject. The papers of our investigators presented here will deal with the research conducted by the Laboratory in the recent years.

Most important for the Geodynamics Project are experimental data on strain (rheology) and strength of rocks. In this connection I would like to present some of the results obtained at our Laboratory.

Several years ago, a special apparatus was developed by I. S. Tomashevskaya, D. B. Balashov and the author, which was capable to exert an axial strain of up to 50 tons on a rock specimen at hydrostatic pressure to 10 kbar [57, 127]. Also, a special experimental procedure was developed for simultaneous measurements of several physical constants of rocks using one specimen (elastic and irreversible deformation, elastic velocities along and across a specimen, strength, microseismicity and electrical conductivity). Another apparatus for pressures to 5 kbar created by O. S. Tomashevskaya, S. M. Kireenkova and the author allows investigation of plastic deformation and strength of rocks in a broad range of strain rates [131].

Strength of rocks was found to increase greatly (in some cases several times or more) under confining pressure. It was established that plastic deformation could be obtained in all types of rocks under high pressure by changing strain rates. It is a common knowledge that rock specimens exhibit various types of failure in pressure tests. All cases can be subdivided into four main types [57, 131]. Theses types are schematically shown in Fig. 1. This figure also shows a curve of acoustic pulse build-up (microseismicity). Differential strain is plotted on the abscissa in per cent of fracture stress σ_s, and a number of pulses n on the ordinate. Type *1* shows plastic flow in a specimen. Microcracks seem to be virtually absent. No acoustic pulses are observed. Types *2* and *3* represent brittle fracture in specimens. The microseismicity curves manifest the formation of microcracks. Type *2* represents shear fracture in which commonly one strong crack is formed at an angle of about 45° to the axis of a specimen. Type *3* is a tension fracture with the formation of one or several cracks parallel to the axis accompanied by a large number of acoustic pulses. Type *4* represents specimen failure with shattering resembling cataclasis. This deformation may be termed quasiplastic. In this case microseismicity is naturally high.

Fracture strain (or strength) of solids is substantially affected by a strain rate. To study this relationship, experiments were performed on rock specimens under hydrostatic pressure at different strain rates. The results of one experiment on granite are shown in Fig. 2. The curves reveal that strength σ_s in a rock is strongly dependent on a strain rate. Decrease in the strain rate causes a decrease in strength. This phenomenon agrees with

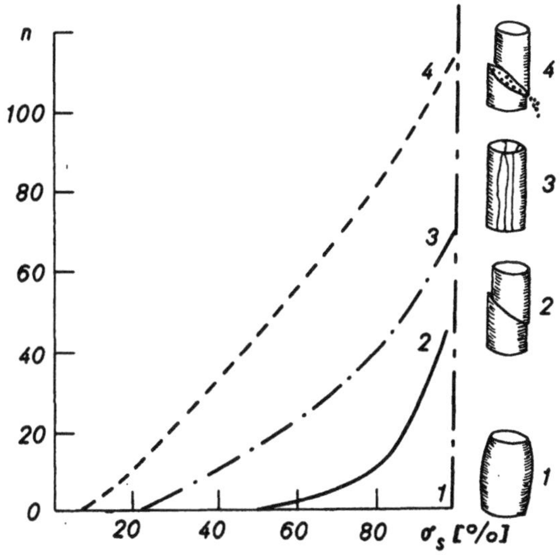

Fig. 1. General types of rocks specimens failure under high pressure and the number of acoustic pulse "n" as a function of differential stress in percent of the strenght.

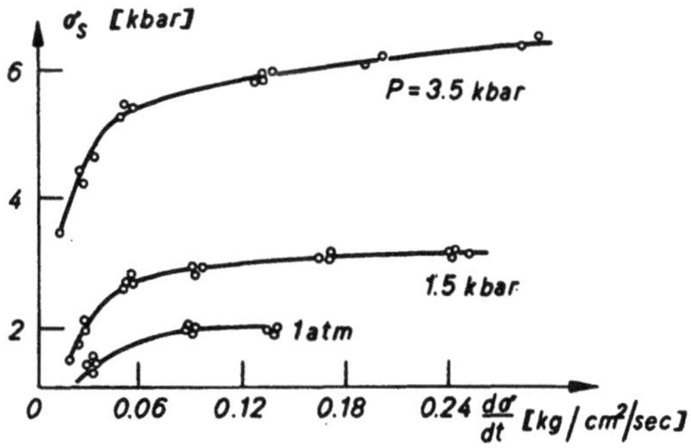

Fig. 2. The strength σ_S for the granite as a function of strain rate σ.

a kinetic theory of strength proposed by S. N. Zhurkov [132]. Figure 2 also shows that strength increases with pressure. At 3.5 kbar the strength of the granite under test was found to be 3 times larger than at atmospheric pressure. This relationship is well illustrated by the curves for two diabase specimens shown in Fig. 3. As seen from the curves, the strength in specimen *1* increased nearly 10-fold at pressure to 8 kbar. This seems to be an effect of healing of microcracks and other defects produced in the rocks by pressure.

Velocities of elastic waves were studied in rocks under high plastic strain by A. I. Levykin, V. A. Budnikov and the author using a matrix chamber of Adams type [133, 134], made of soft steel, in which a rock specimen is compressed by two pistons through the action of a press. The chamber was improved. The improvements were as follows. The matrix was made sectional in design. Specimens were placed in thin lead jackets. Steel rings of triangle cross-section were incorporated for packing on the pistons. Ultrasonic transducer and detector were attached to the ends of the pistons. It is evident that the strain of a specimen in our case is nonuniform. Fig. 4 shows a stress curve for granite. The curve

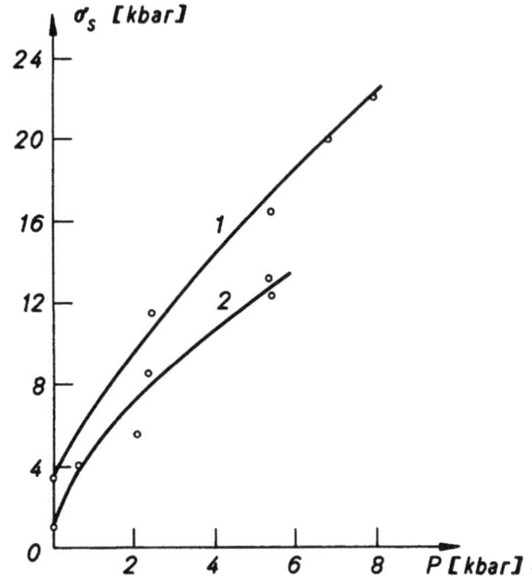

Fig. 3. The strength σ_S versus the pressure p for two diabase specimens.

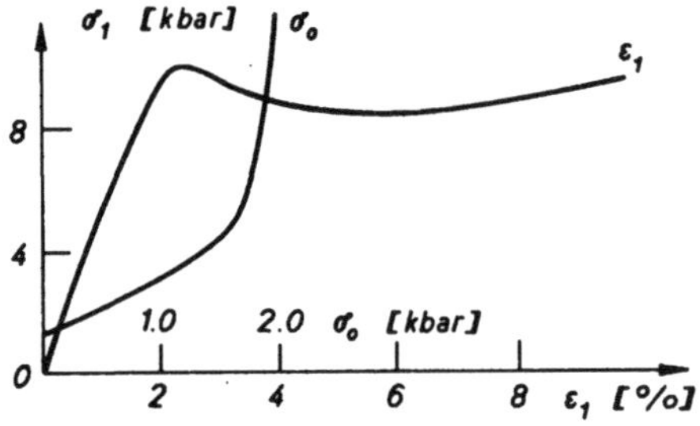

Fig. 4. Lateral stress σ_0, and also axial deformation ε_1 as a function of axial stress σ_1 for granite.

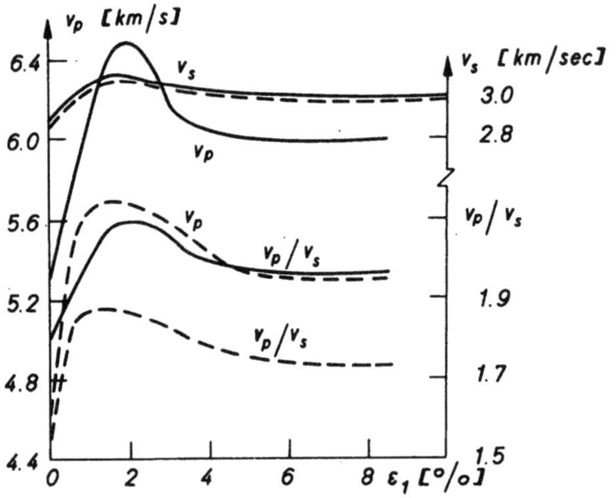

Fig. 5. P-wave, S-wave velocities and their ratio versus axial deformation ε_1 for dry granite specimen (dotted line) and water saturated specimen (continuous line).

is a lateral stress σ_0 vs. axial stress σ_1 relationship. The figure also shows a stress-strain curve, i. e. ε_1 as a function of σ_1. The compression of a specimen is seen to occur at a variable load parameter, $d\sigma_0/d\sigma_1$, which is substantially larger at the initial stage of the test. The stress-strain curve depicts a well-seen flow point. As a result, brittle rocks (granite, gneiss, gabbro, etc.) acquire plastic properties. At the end of a test, specimens have a shape of a barrel with residual deformation amounting to 10 per cent or more.

Fig. 5 shows the results of ultrasonic measurements in granite. Dotted lines indicate the dry tests and solid lines the tests with interstitial water. Both in dry and water saturated specimens, velocities increase in the regular way in the elastic region up to a flow point, whereas in the region of nonelastic strain velocities decrease. Variation of compressional velocities is 2—3 times as great as that of shear velocities. In accordance with this relationship the V_p/V_s ratio also increases to a flow point and then decreases.

Some problems of the Geodynamics Project require experimental data on viscosity of rocks, which reflects the processes of their plastic flow. As a result of the experiments performed by I. S. Tomashevskaya, S. M. Kireenkova, J. N. Khamidullin and the author, using the above mentioned apparatus, it was possible to determine the effective viscosity of granite, granodiorite, gabbro and limestone [135]. The viscosity, η, of the rocks was found to be greatly dependent on a strain rate or gradient $d\varepsilon/dt$. The relationship thus obtained is shown in Fig. 6 in logarithmic scale. Solid circles show the results obtained for the above mentioned rocks under pressure of 2—3 kbar and open circles the results of other investigators for similar rocks [95]. It is remarkable that the viscosity values in all the tested rocks are rather close at a certain strain rate. Only in gabbro they are somewhat larger. Besides, it should be noted that in parallel experiments viscosity is greatly variable in one and the same rock. Viscosity variation with pressure to several kilobars was found to be small as compared to that with a strain rate.

In Fig. 6 our measurements of viscosity obtained at the strain rates of 10^{-4} to 10^{-7} sec^{-1} are 10^{12} to 10^{14} poise. The results of other investigators measured at the indicated strain rates fit into the same interval. In general, all the solid and open circles in Fig. 6 are seen to group near one straight line. It is remarkable that the viscosity values of 10^{19}—10^{22} poise obtained from theoretical geodynamic calculations also show a good fit to this line. In further rheologic investigations we should accumulate experimental data on viscosity of rocks at various strain rates in order to fill a gap existing in the range of $d\varepsilon/dt = 10^{-8}$—$10^{-18}$ sec^{-1}, and to clear up the difference in viscosity for different types of rocks.

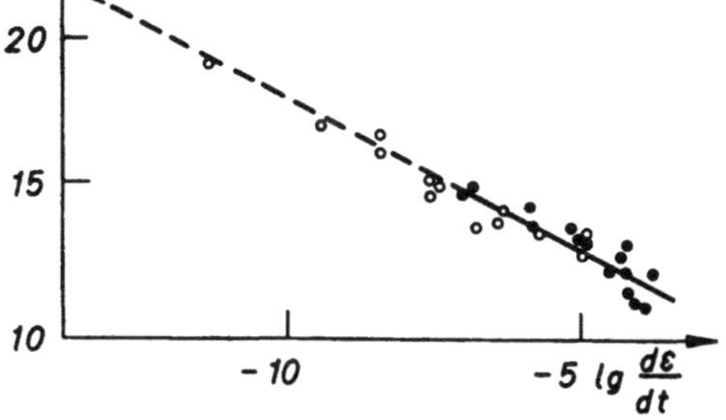

Fig. 6. Dependence of effective viscosity on shear velocity $\dot{\varepsilon}$.

Geodynamics, in particular, tectonics, requires the knowledge of the structure and composition of the Earth's crust and upper mantle. The experimental data available on velocities and densities of rocks under high pressures allow us to predict the presence of certain real rocks in layers and blocks of the Earth's interior in accordance with field geophysical and geological observations. Several examples of such predictions are reported in the literature [57, 136].

Fig. 7 shows a seismic section across the south-western part of Armenia, plotted from the data of [137]. A pronounced feature in the figure is a layer of low compressional velocity above the Moho discontinuity, at a depth of 40—50 km. A supposition was expressed that this layer might be composed of serpentinized ultrabasic rocks. The experimental study of the rock samples collected from this region showed that the metamorphosed ultrabasic rocks, serpentinites, had low density and low velocities of compressional and shear waves throughout the test pressure to 20 kbar [137]. The lowest velocities were found

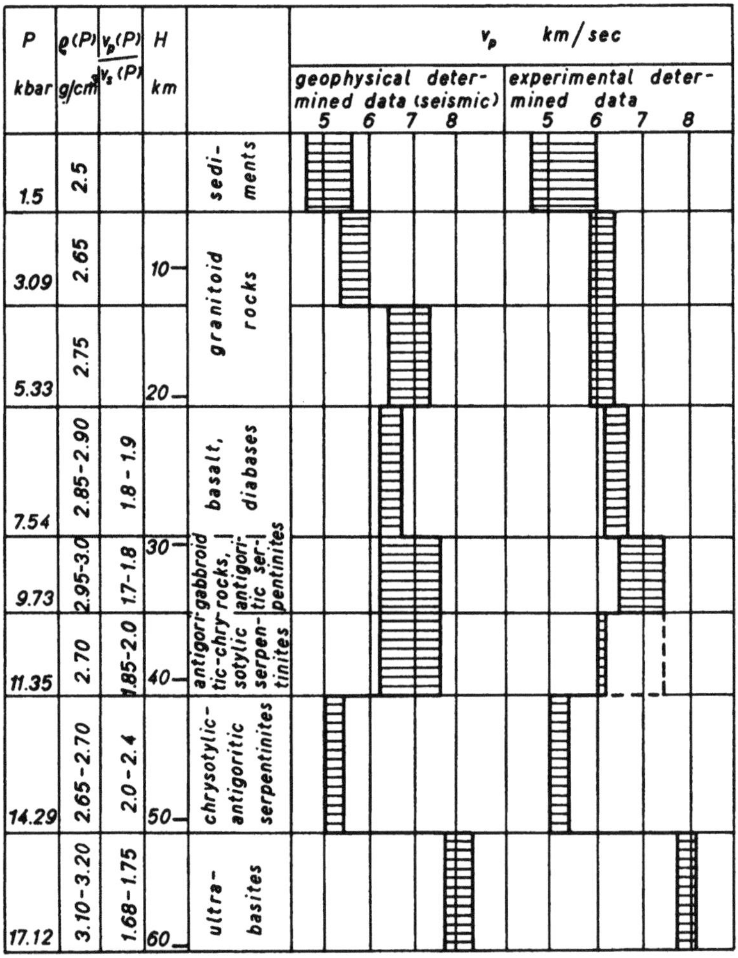

Fig. 7. Seismic section of the Earth crust across the south-western part of Armenia. Left side-seismic data, right side — laboratory data.

in serpentinites containing 70% chrysotile, namely $V_p = 5$ km/sec, at 15 kbar. Similar results were previously obtained for serpentinites from the Voronezh crystalline massif, among which lizardite serpentinites exhibited extremely low velocities (curve *4*) [138]. Curve *1* is shown for unaltered periodotite which naturally has a high compressional velocity. Curves *2* and *3* were obtained for antigorite serpentinites.

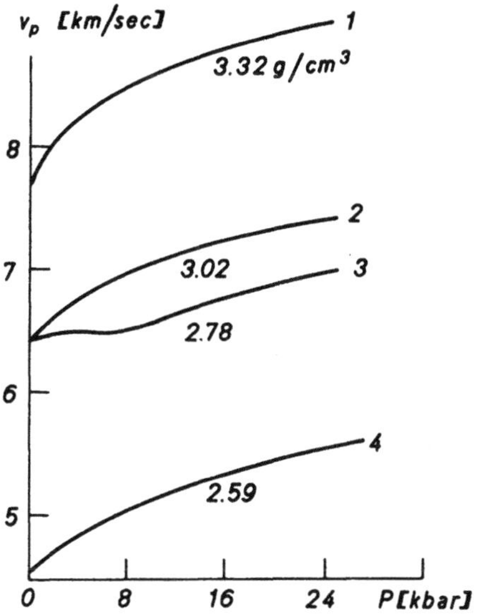

Fig. 8. Dependence of *P*-wave velocities on the pressure *p* for the rocks from the Voronezh crystalline massif. *1* — peridotite, *2*, *3* — antigorite serpentinites, *4* — lizardite serpentinite.

Let us compare the two crustal velocity structures given in Fig. 7: the seismic section on the left hand, and the velocity distribution plotted from experimental mesurements on the rocks of the same region of Armenia made at pressures and temperatures which are believed to exist at certain depths. The latter characterized an inferred petrophysical composition of the lithosphere. It is seen that to a depth of 40 km compressional velocities in the crust gradually increase, and the above mentioned low-velocity layer corresponds to chrysotile-enriched serpentinites. It is worth noting that serpentinites possess a high plasticity and are, in this connection, of interest to geodynamic investigations.

RESIDUAL STRAIN IN GRANITES UNDER DIFFERENTIAL STRESS

L. I. Zviagintsev, I. S. Tomashevskaya

Granites that loock indentical in outcrops are under microscope identified as deformed if residual strain is present in minerals, and as undeformed if it is absent [139, 140]. Moreover deformed granites are classified into varieties that underwent uniform flow or shear strain. These three granite varieties differ in average parameters determined at ordinary atmospheric conditions. This is illustrated by the table below which shows the results for Caucasus granites occurring at moderate depth (3—4 km).

Table 1. Average physical parameters of mesoabyssal granites

Location. Deformation type	Number of samples	Density gm/cm³	Effective porosity, %	Velocity V_p km/sec	Velocity V_s km/sec	V_p/V_s
Dashkesan intrusion Undeformed	20	2.60	2.09	5.07	3.27	1.58
Megrin intrusion Uniform flow	21	2.62	1.41	4.8	2.83	1.69
Eljurtin intrusion Shear strain	27	2.60	1.74	3.3	2.27	1.44

The table shows that elastic velocities V_p and V_s are lower in deformed granites as compared to undeformed ones. The V_p/V_s ratio is high for the rocks that underwent uniform flow and low for the rocks which suffered shear strain.

Deformed granites exhibit reduction of porosity in spite of the appearance of shear chracks and even extension cracks. This shows that compaction under overburden pressure occurs at the expense of porosity. It may be assumed that residual deformation in granite arises after crystallization as a result of stress exceeding lithostatic load.

To check this assumption experiments were performed using specimens of granite formed at intermediate depth (mesoabyssal facies). The specimens were undeformed rocks having the initial bulk porosity of 2.1 per cent and effective porosity of 1.8 per cent. The specimens, 20 by 50 mm in size, were subjected to hydrostatic pressure of 1.5 kb which corresponds to the overburden load of 4—5 km. Additionally, differential stress was exerted on both ends of specimens at a rate of 0.05 kb/sec. Specimen failure occured at a differential stress (σ_{max}) of 5.4—5.5 kb. Differential stress (σ), compression ($\varepsilon_{||}$) and shear ($\varepsilon_{\perp}$) strains, as well as acoustic pulses (n_s) in a range of up to 3 kcps were continuously recorded during the experiments. The experimental test could be stopped at any deformation stage or any compressive stress lower than the ultimate fracture stress (σ_{max}). Porosity was determined in the deformed specimens and residual deformation was measured in thin sections. Acoustic pulses were recorded in granite specimens from the beginning of the differential stress application (Fig. 1). The most violent departure of the deformation from being linear was found to coincide with an abrupt increase of acoustic pulses. This occurred in the region close to the fracture stress, exceeding 80—95 per cent of σ_{max}. Specimen failure was brittle in all the experiments. The longitudinal strain ($\varepsilon_{\|}$) reached 0.35—0.95 per cent and the transverse strain ($\varepsilon_{\perp}$) did not exceed 0.5 per cent. What is the cause of the non-uniform strain and of the indicated behaviour in granite under differential stress?

Prior to be experimentally deformed, granite specimens did not exhibit residual deformation (Fig 2a). Yet, at the beginning of a compression test (25—30 per cent of σ_{max}),

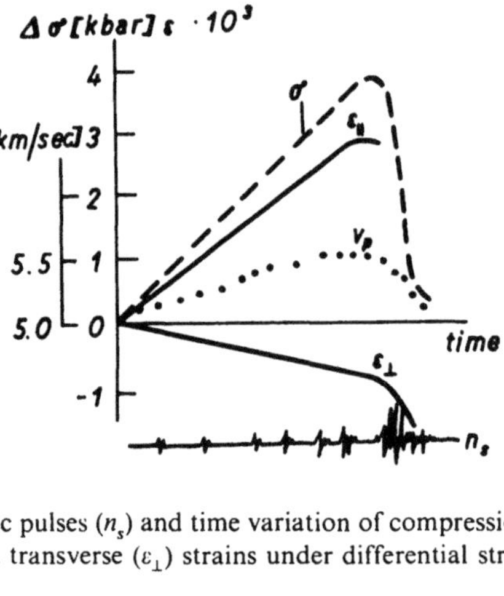

Fig. 1. The distribution of acoustic pulses (n_s) and time variation of compressional velocity V_p, longitudinal ($\varepsilon_{\|}$) and transverse ($\varepsilon_{\perp}$) strains under differential stress.

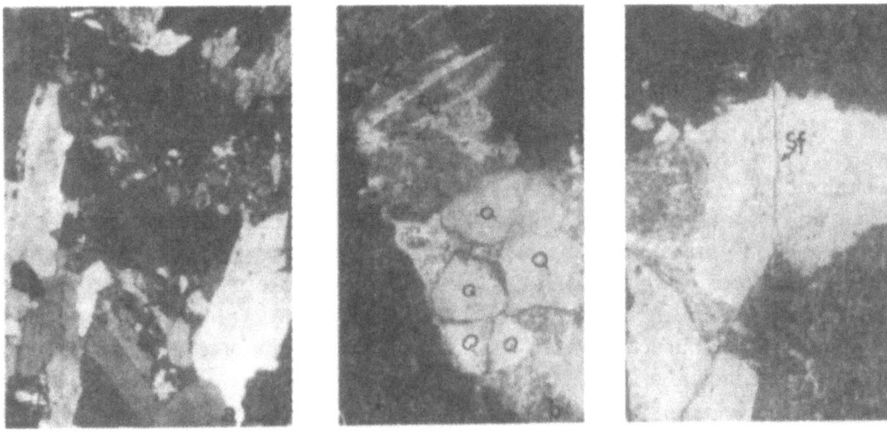

Fig. 2. Photomicrographs of granite thin sections
a) undeformed, b), c) deformed at P = 1.5 kb and 25 per cent of σ_{max}, t_s – translation in plagioclase crystal (F^l), Q – shattered quartz, sf – shear fracture.

effects of deformation were visible in thin sections. The principal mechanism of deformation of plagioclase crystals was found to be {010}-translation in one of twinning lamellae (Fig. 2b). Microcline exhibited twin gliding. Gliding was also noticed in quartz, in which point defects (fluid inclusions) slipped under stress, being arranged as sub-parallel chains in the zone of the greatest shear stress. In places the defects coalesced to form cracks along which gliding occurred. Cataclastic flow was widely developed. Quartz grains were separated in to blocks and the interstices between them were filled with fine-grained felspar material (Fig. 2b). Intergranular displacements occurred at contacts of grains having different elastic properties, for instance in quartz and microcline grains. Shear cracks were formed at this stress, which originated owing to the concentration of stress in some limited areas like, for instance, in the places of face edges pressing on the adjacent plane of a mineral (Fig. 2c). In some cases shear cracks were formed at cleavage planes. Extension cracks were very rare. They were formed within minerals. Macroscopically they were not visible.

The amount of brittle fracture increased at differential stress of about 50 per cent of σ_{max}. In this case twinning developed on a broader scale. Very common were wedge twins with accomodation zones. Twin growth was restrained by various disruptions. Shear cracks exhibited a tendency to group in a zone inclined 30—35° to the axis of compression. Macroscopically, the specimens remained monolithic.

At differential stress of 75 per cent of σ_{max}, the general structural appearance of granite was still preserved. Recrystallization was observed in microcline, plagioclase (Fig. 3a) and even in hornblende minerals. In places the random structure was locally converted to the oriented one. Crystals were rotated so that their long sides were perpendicular to a shear plane. Shears were localized in an eventual fracture zone (Fig. 3b). Macroscopically, the specimens were still monolithic.

At differential stress exceeding 75 per cent of σ_{max} began an process of intense growth and coalescence of cracks in planes of the greatest shear stress. This process was recorded by a great increase of acoustic pulses (Fig. 1), which implies growth of brittle fracture.

The fractured specimens exhibited one main disruption, which was a cataclastic shear crack. It was filled with fragments and friction sand. Minor shear and tension cracks

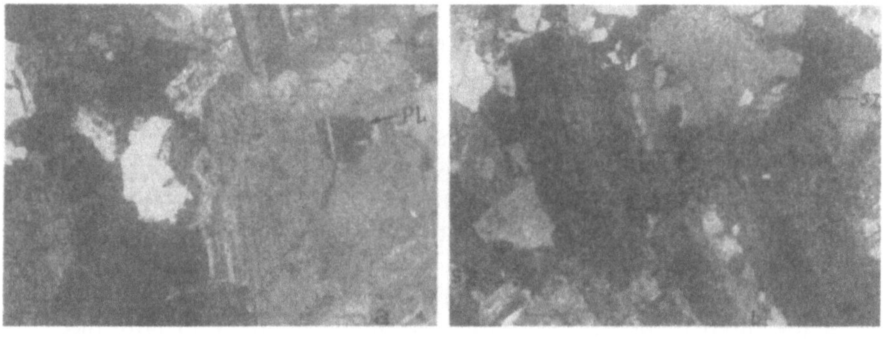

Fig. 3. Photomicrographs of granite deformed at $P = 1.5$ kb and 75 per cent of σ_{max}
a) recrystallization in plagioclase, a new crystal (*Pl*) is formed at the end of a fracture; b) orientation of crystal with long sides perpendicular to shear zone (*sz*).

branched off from the main one. All the cracks were inclined at an angle of 30—35° to the axis of compression.

Two regions were observed in the process of deformation, which were distinguished by a number of cracks. A boundary between them was a zone of the intense formation of disruptions, which began at a stress exceeding 80 per cent of σ_{max}.

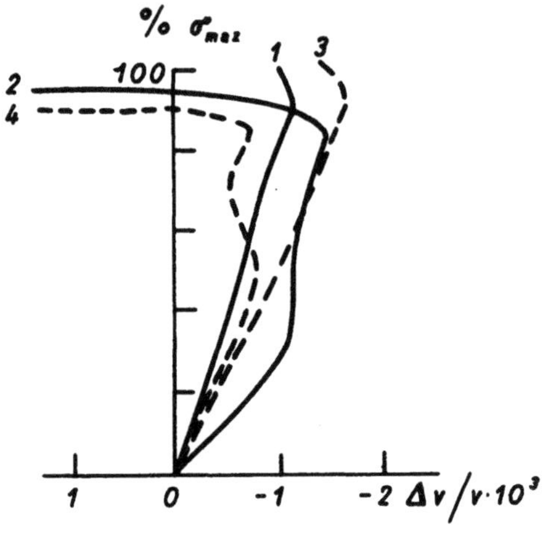

Fig. 4. Average volume change ($\Delta V/V$) as a function of differential stress (σ_{max}). *1*, *2* dry rocks; *3*, *4* rocks partially saturated with water.

One of the regions exhibited uniform flow resulting from plastic deformation in minerals. In this case compaction of the rock increased, which was evidenced by a decreased porosity. In granite specimens of 2.07% bulk and 1.78% effective porosity, subjected to dry compression, the former decreased to 1.61—1.80% and the latter to 1.39—1.48%. The value of effective porosity was close to the average porosity (1.41%) inherent to granites which underwent uniform flow in natural conditions. In the deformed rocks partially saturated with water, effective porosity was still lower (1.15—1.20%). Compaction was recognizable in a relative reduction of volume (Fig. 4) which was calculated using the measured average longitudinal ($\varepsilon_{\parallel}$) and transverse ($\varepsilon_{\perp}$) deformations by the equations:

$$\Delta V/V = -\varepsilon_{\parallel} + 2\varepsilon_{\perp} .$$

The second region was distinguished by the intense formation of cracks, which resulted in separation of a specimen into blocks. At this time cohesion reduced owing to the cracks formed in the narrow region, and porosity increased twofold. The loss of cohesion was accompanied by a relative enlargement of the volume (Fig. 4). Yet, the unbroken blocks were still more compact as compared with the initial granite.

There are two mechanisms leading to the breakdown of a rock. These are plastic flow and brittle fracture. Both occur simultaneously. Potential possibility of both mechanisms is inherent to the rock itself owing to the anisotropy of elastic properties of minerals and polycrystalline nature of the rock. Contact planes do not appear to be sources of low strength.

Obviously, those natural granites which exhibit a deformation of uniform flow underwent compaction at non-uniform compression stress. Whereas the granites in which shear fractures are present seem to have suffered non-uniform compression at axial stress close to the critical one at a given overburden pressure.

The comparison of the experimental results and natural observations shows that residual deformations in the rocks reflect the geodynamic conditions under which granite intrusions were formed beginning from the earliest phases of melt crystallization. Residual deformations combined with physical properties of rocks allow the solution of an inverse problem, i. e. the reconstruction of the geodynamic regime of the emplacement of granite intrusions.

SYSTEM OF CRACKS IN A STRESS FIELD

J. NIEWIADOMSKI

The paper gives a method of calculating the distribution of stresses and displacements in an isotropic body containing a finite number of "sliding mode" [142] cracks. We assume that the cracks are parallel to each other. A two-dimensional problem is considered. System of such type occour in natural conditions. They may be caused by the action of unidirectional, nonhydrostatic outer stresses or by geological conditions.

The stresses will be described by a tensor with components δ_x, δ_y and τ_{xy}, the displacements by a vector $u(u_x, u_y)$. If a crack lies along the x-axis of a rectangular coordinate system (x, y), the boundary conditions existing on this crack can be expressed as

$$\begin{aligned} &\text{a)}\ [\delta_y]^s = 0; \\ &\text{b)}\ \tau_{xy} = 0; \qquad [\tau_{xy}]^s = 0; \\ &\text{c)}\ [u_{y,x}]^s = 0; \quad u_y(x) \in C^1 . \end{aligned} \tag{1}$$

Symbol $[\cdot]^s$ means a jump of the quantity in the brackets, which occurs on the crack under consideration. C^1 is a set of continuos functions.

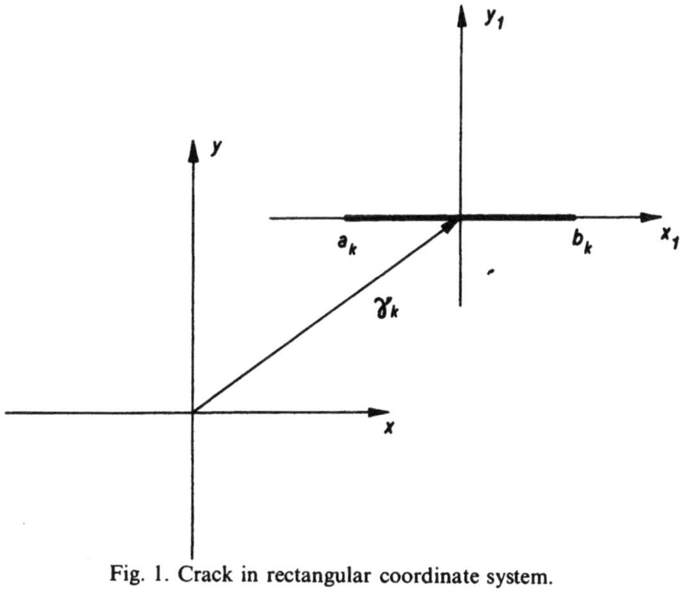

Fig. 1. Crack in rectangular coordinate system.

Let a single crack lie along the x_1-axis of system (x_1, y_1). We will be interested in stresses in the system (x, y), obtained by a parallel displacement by vector γ relative to the previous system (x_1, y_1) (Fig. 1). Making use of basic formulae of theory of elasticity [143], we can write the following expression for the stress and displacement components in these two coordinate systems:

$$\begin{aligned} \delta_x + \delta_y &= \delta_{x_1} + \delta_{y_1}, \\ -\delta_{x_1} + \delta_{y_1} + 2i\tau_{x_1 y_1} &= -\delta_x + \delta_y + 2i\tau_{xy}, \\ 2\mu(u_x - iu_y) &= 2\mu(u_{x_1} - iu_{y_1}), \end{aligned} \tag{2}$$

μ — the Lamé elastic parameter.

Let $\Phi(z)$ and $\Psi(z)$ be the Muskhelishvili potentials in system (x, y), and $\Phi_1(z_1)$ and $\Psi_1(z_1)$ be the potentials in system (x_1, y_1). Let z and z_1 be complex variables: $z = x + iy$, $z_1 = x_1 + iy_1$, $i = \sqrt{-1}$. Then [143]

$$\begin{aligned} &\text{a)} \quad \delta_{x_1} + \delta_{y_1} = 4\, Re\, \Phi(z), \\ &\text{b)} \quad \delta_{y_1} - \delta_{x_1} + 2i\tau_{x_1 y_1} = 2(\bar{z}\Phi'(z) + \Psi(z)), \\ &\text{c)} \quad 2\mu(u_{x_1} - iu_{y_1}) = \varkappa\overline{\varphi(z)} - \bar{z}\varphi'(z) - \psi(z), \end{aligned} \tag{3}$$

$$\frac{d\varphi(z)}{dz} = \Phi(z); \qquad \frac{d\psi(z)}{dz} = \Psi(z); \qquad \varkappa = \frac{\lambda + 3\mu}{\lambda + \mu};$$

λ — the Lamé elastic parameter.

Differentiating equation (3, c) with respect to variable z in the x-axis direction, we obtain

$$2\mu(u_{x_1, x_1} - iu_{y_1, x_1}) = -\varkappa\overline{\Phi(z)} + \Phi(z) - \bar{z}\Phi(z) - \Psi(z). \tag{4}$$

From condition (1, a) we have

$$\begin{aligned} Im\, [-\bar{z}\Phi'(z) - \Psi(z)]^s &= 0; \\ Im\, [-\varkappa\Phi(z) + \Phi(z) - \bar{z}\Phi'(z) - \Psi(z)]^s &= 0. \end{aligned} \tag{5}$$

It follows from the above two formulae that on the crack there is:

$$Im\, [\Phi(z)]^s = 0. \tag{6}$$

Potential $\Phi(z)$ can be expressed by function $h(t)$, $Im\,(h(t)) = 0$, as a Cauchy integral

$$\Phi(z) = \frac{1}{2\pi i}\int_a^b \frac{h(t)}{t - z}\, dt + \Phi(\infty), \tag{7}$$

where a and b are coordinates of the beginning and the end of the crack.

Let us introduce a potential $\Omega(z)$:

$$\Omega(z) \equiv 2\Phi(z) + z\Phi'(z) + \Psi(z). \tag{8}$$

If the crack lies along the axis of the (x, y) coordinate system, then it follows from the boundary conditions for the crack that $\Omega(z)$ has equal values on the entire plane, i.e.

$$\Omega(z) = \Omega(\infty) = \text{const}. \tag{9}$$

This value is determined by the outer stresses. However, if the crack lies along the x_1 axis of system (x_1, y_1), displaced by vector γ relative to the system (x, y), then — making use of conditions (2) and (3) — we obtain:

$$\bar{z}_1 \Phi_1'(z_1) + \Psi_1(z_1) = \bar{z}\Phi'(z) + \Psi(z) . \quad (10)$$

Upon introducing potential $\Omega(z)$ the above equation can be written in the form

$$\bar{z}\Phi_1'(z_1) + \Omega_1(z_1) - 2\Phi_1(z_1) - z_1\Phi_1'(z_1) = \bar{z}\Phi'(z) + \Omega(z) - 2\Phi(z) - z\Phi'(z) . \quad (11)$$

Taking into ccount that $z = y + z_1$ and that $\Phi_1(z_1) = \Phi(z)$, we can obtain the following expression for function $\Omega(z)$:

$$\Omega(z) = \Omega(\infty) + \frac{(\gamma - \bar{\gamma})}{2\pi i} \int_{a_1}^{b_1} \frac{h(t)}{(t - z)^2} dt . \quad (12)$$

Potentials Φ and Ω enable us to calculate the stresses and displacements in each point of the body [144],

$$\begin{aligned} \delta_x(z) &= -Re\,\{(z - z)\,\Phi'(z) + \Omega(z)\} + 4\,Re\,\Phi(z); \\ \delta_y(z) &= Re\,\{(z - z)\,\Phi'(z) + \Omega(z)\}; \\ \tau_{xy}(z) &= Im\,\{(z - z)\,\Phi'(z) + \Omega(z)\} - 2\,Im\,\Phi(z) . \end{aligned} \quad (13)$$

Consider system of q cracks, defined in coordinate system (x, y) by the coordinates of beginning and end of each crack, a_k and b_k respectively. For such a system potentials $\Phi(z)$ and $\Omega(z)$ can be written as the following sums

$$\Phi(z) = \frac{1}{2\pi i} \sum_{k=1}^{q} \int_{a_k}^{b_k} \frac{h_k(t)}{t - z} dt + \Phi(\infty); \quad (14)$$

$$\Omega(z) = \frac{1}{2\pi i} \sum_{k=1}^{q} (\gamma_k - \bar{\gamma}_k) \int_{a_k}^{b_k} \frac{h_k(t)}{(t - z)^2} dt + \Omega(\infty); \quad (15)$$

$$\gamma_k = 0.5\,(a_k + b_k).$$

Function $h_k(t)$ should be determined with the allowance for the boundary conditions on the cracks and the outer stresses. As follows from the boundary conditions on the s-th crack,

$$Im\,\{(\bar{z}_s - z_s)\,\Phi'(z_s) + \Omega(z_s) - 2\Phi(z_s)\} = 0; \quad (16)$$
$$z_s \in \langle a_s, b_s \rangle .$$

In an expanded form, this condition becomes

$$\frac{1}{\pi} \int_{a_s}^{b_s} \frac{h_s(t)}{t - z_s} dt + \frac{1}{\pi} \sum_{\substack{k=1 \\ k \neq s}}^{q} \int_{a_k}^{b_k} h_k(t)\,\{Re(t + \gamma_k - \gamma_s - z_s)^{-1} + \\ + Im(\gamma_k - \gamma_s)\,Im(t + \gamma_k - \gamma_s - z_s)^{-2}\}\,dt = -Im\,\Omega(\infty). \quad (17)$$

Equations of this type, written for all q cracks, provide an initial system of equations for calculating the unknown functions $h_s(t)$. Each of these functions can be expressed by a function $v_s(t)$, which is smooth on segment $\langle a_s, b_s \rangle$

$$h_s(t) = v_s(t) \{(t - a_s)(b_s - t)\}^{-1/2} . \tag{18}$$

The system of equation (17) can be solved numerically, making use of function $v(t)$ and the quadrature formulae given by [145].

Let l_s be the s-th crack length. The boundary condition (1) should be satisfied on each crack. This is equivalent to the condition (16).

Writing this condition for all q cracks and making use of the analyticity of function Φ, Ω in the area outside the cracks, we obtain the following system of algebraic equations which substitute system (17)

$$\sum_{\substack{k=1 \\ k \neq s}}^{q} \sum_{m=1}^{n} v_k(x_m) \{Im(\gamma_k - \gamma_s)\, Im(\gamma_k - \delta_s + l_k x_m - l_s y_k)^{-2} + Re(\gamma_k - \gamma_s + l_k x_m - l_s y_k)^{-1}\} = -n\, Im\, \Omega(\infty) ; \tag{19}$$

$$\sum_{m=1}^{n} v_s(x_m) = 0 ;$$

$$x_m = \cos \frac{(2m-1)}{2n} \pi ; \qquad y_k = \cos \frac{k\pi}{n} ; \qquad m = 1, 2, \ldots, n ;$$

$$k = 1, 2, \ldots, n-1 , \qquad s = 1, 2, \ldots, q .$$

We thus have a system of equations for $n \cdot q$ unknown values of $v(x)$. For example, for a system of three cracks, each divided into $n = 20$ parts, we obtain a system of 60 linear equations. The reason for dividing the i-th crack into parts with coordinates $\gamma_i + l_i \cos \frac{2m-1}{2n} \pi$ is to reduce the number of mathematical operations in solving system of equations (17). Moreover, this division is convenient due to the fact that it becomes denser around the ends of the cracks, where the stress and displacement gradients are the greatest. Comprehensive information on the transformation of integral equations (17) into a system of linear equation (19) can be found in papers [144, 146].

The values of potentials $\Phi(z)$ and $\Omega(z)$ in the "infinity" determine the so-called principal stresses of the outer stress field [147]. If an angle between the x-axis and the direction of the principal stress N_1, N_2 is α, and $\alpha \in \langle -\pi/2, \pi/2)$, then

$$\begin{aligned} \Phi(\infty) &= 0.25(N_1 + N_2) \\ \Omega(\infty) &= 0.5(N_1 + N_2) - (N_1 - N_2)\, e^{-2i\alpha} . \end{aligned} \tag{20}$$

The solution of above problem, i. e. the calculation of stress and displacement for a system of cracks, enables us to solve several important questions concerning, for example, the mechanics of destruction. One can mention here the energy of mutual interaction of cracks in the outer stress field or the application of strength hypotheses to a system of cracks.

CHAPTER III

Electrical Properties

RESISTIVITY OF MINERALS AND ROCKS OF THE UPPER MANTLE AS RELATED TO GEODYNAMIC PROCESSES

E. I. PARKHOMENKO

The present-day knowledge suggests that the Earth's crust and part of the upper mantle are dynamically most active parts of the Earth. Various processes of material transformation occur in these regions. These are dehydration, decarbonatization, chemical reactions of decomposition and replacement, polymorphism and partial or complete melting. Transformation of the material involves changes in the physical properties of rocks, including electrical ones [148]. In this connection an important step in studying the physical mechanism of abyssal processes is experimental investigation of electrical properties of minerals and rocks along with the other physical properties under high pressure and high temperature.

Besides, information on the resistivity of rock-forming minerals of the mantle material and of an ultrabasic rock assemblage is of independent interest, being highly valuable for further development of the theoretical studies of the Earth's interior.

If we proceed from pyrolite composition of the upper mantle, the most probable constituents of its material are olivine, pyroxenes and garnets. With any association of the indicated minerals, the following oxides must prevail in the chemical composition of the upper mantle material: SiO_2, MgO, FeO, Al_2O_3 and CaO. In this connection let us consider first of all the electrical parameters of these oxides, most of which occur in nature as monomineral formations.

Minerals, such as quartz, SiO_2, periclase, MgO, and corundum, Al_2O_3, are characterized by the highest values of resistivity in a wide range of temperatures (Table 1). Only bromellite, BeO, stands in one row with these minerals.

Table 1. Resistivities of oxide minerals and cation parameters

Oxide	ϱ, [ohm. cm] at t [°C]		Energy of crystal lattice, [eV] (per a pair of ions)	Cation	Ionic radius, r [10^{-8} cm]	Polarization capacity, d [10^{24} cm^3]
	600	1000				
BeO	10^{10}	10^8	—	Be^{2+}	0.34	—
SiO_2	10^8	10^6	—	Si^{4+}	0.39	0.039
Al_2O_3	10^{10}	10^9	157.0	Al^{3+}	0.57	0.067
MgO	10^{10}	10^9	41.36	Mg^{2+}	0.76	0.144
CaO	10^8	10^6	37.53	Ca^{2+}	1.06	0.531

High insulation capacity of the minerals concerned is explained by the following. These minerals are predominantly characterized by ionic bonds and a high energy of crystal lattice. They are formed of bivalent, trivalent and tetravalent cations which have small ionic radia and low polarizability. As a result, the energy necessary for an ion to be interstitially displaced from a normal state or to occupy a vacant lattice point is substantially greater than in the case of univalent cations of alkali metals.

Thus, conductivity of minerals formed of identical cations must be largely determined by a cation type. The same is clearly recognizable in the case of silicates.

An olivine group includes forsterite, fayalite and olivine proper. The largest resistivity is found in forsterite which is a pure magnesia member in the mineral series concerned [149, 150].

Olivine, $(Mg, Fe)_2 SiO_4$, a ferriferous variety of forsterite in a form of a solid solution, has a lower resistivity. A broad range of ϱ values is established for ionic crystals and polycrystal derivatives of olivine in the region of impurity conductivity at a temperature of 200—300 °C which excludes thermochemical reactions. This diversity of ϱ values is due to variable FeO and Fe_2O_3 contents and various ratios Fe^{2+}/Fe^{3+}. Besides, it is found that olivines containing Fe^{3+} have an activation energy of less than 1 eV. In the cases when iron is present in Fe^{2+} an activation energy and resistivity are higher [150]. The deciding role of Fe^{3+} is explained by the fact that in the simultaneous presence of ferric oxide, Fe_2O_3, and ferrous oxide, FeO, electric conductivity is accomplished by the transition of an electron from the iron Fe^{2+} to the adjacent ion Fe^{3+} at a low activation energy, which contributes substantially to the conductivity of olivines.

Fayalite, Fe_2SiO_4, is distinguished by a large amount of FeO (up to 76 per cent). In addition, Fe_2O_3 is frequently present. For this reason, the resistivity, E_0 and ϱ_0 of fayalite are essentially lower than the same parameters of olivine.

The most common minerals in the garnet group are pyrope and almandine. Like fayalite, almandine containing 43.3% FeO exhibits a smaller resistivity than pyrope which does not contain iron oxides. It is remarkable that almandine usually yields a higher resistivity than aegirine though it is much richer in FeO than the latter. One of the reasons of the inconsistency between ϱ and the number of Fe cations seems to be a larger Fe^{2+} coordination number in almandine, $n = 8$, as compared to olivine ($n = 6$).

The final group of minerals which plays an important role in the composition of the upper mantle consists of pyroxenes: enstatite, chrome-diopside, diopside, jadeite, hedenbergite, spodumene and aegirine. In spite of the identical structure, the minerals differ in resistivity, activation energy and resistivity coefficient (Fig. 1). This group of minerals is a good illustration of a role of a cation type in the formation of resistivity. Variation in the cation composition in an enstatite-aegirine series, which consists in the replacement of the bivalent cations Mg^{2+}, Ca^{2+} by the univalent cations Li^{1+}, Na^{1+}, and in the growth of Fe cations, is responsible for a broad rangs of ϱ values. As a result, at $t = 200$ °C the ϱ values decreases from 10^{12} ohm · cm in enstatite to 10^5 ohm · cm in aegirine. The difference by six orders between the ϱ values in the two extreme minerals of this group holds at $t = 600$ °C. Only at $t = 1000$ °C, it becomes smaller decreasing to four orders (10^6 ohm · cm and 10^2 ohm · cm, respectively).

Minerals containing Fe-cations exhibit a decrease of resistivity in a temperature range of 200—650 °C and at quasihydrostatic pressure to 20 kbar. And the most intensive decrease of resistivity with pressure is observed in almandine and aegirine (Fig. 2). A high intensity of resistivity variation with pressure in ferriferous minerals can be explained by the

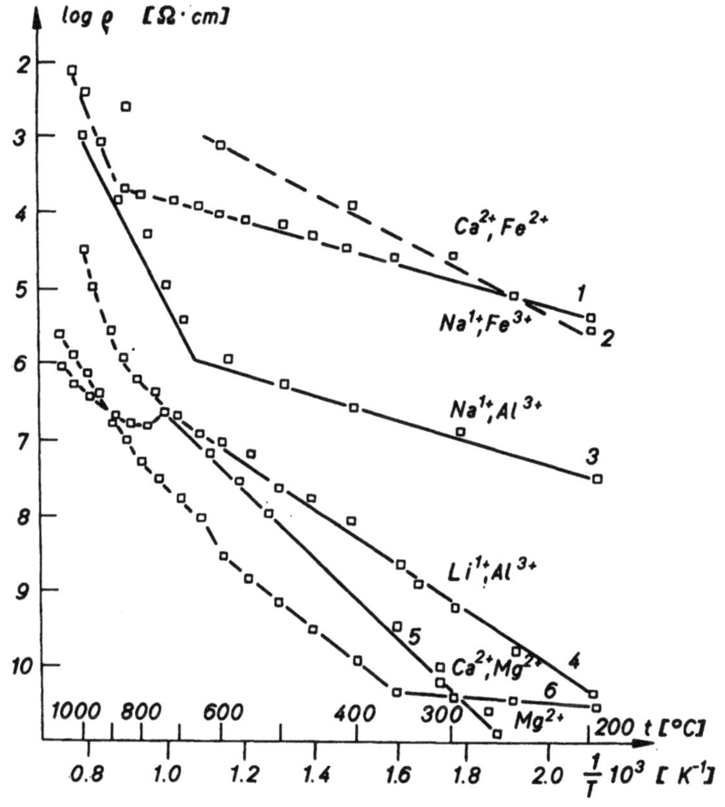

Fig. 1. Temperature dependence of the resistivity of pyroxenes: *1* — aegirine; *2* — hedenbergite; *3* — jadeite; *4*. — spodumene; *5* — diopside; *6* — enstatite.

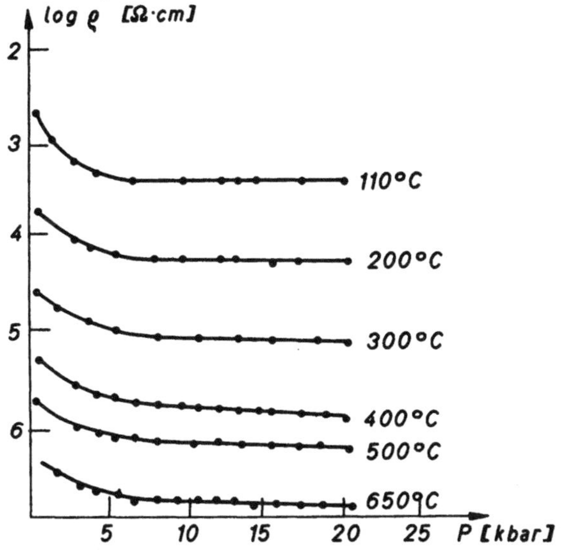

Fig. 2. Isothermes of ϱ for aegirine under pressure 1—20 kbar.

intensification of conductivity mechanism through the convergence of Fe ions in the process of deformation, which promotes a decline in the activation energy of the exchange $Fe^{2+} - e \rightleftarrows Fe^{3+}$.

A different relationship is observed in minerals which do not contain Fe cations in their chemical formulas, and in which iron is not present as impurity. Resistivity in such minerals increases as quasihydrostatic pressure grows. This relationship is observed in jadeite, spodumene and forsterite. Apparently, this behaviour is conditioned by the ionic nature of conductivity in these minerals.

Experimental data at higher pressures and temperatures were primarily obtained for minerals of the olivine group. The data available reveal a weak dependence of resistivity on pressure [151]. Activation energy in fayalite and olivine decreases with pressure, in forsterite increases.

Owing to the fact that olivines contain Fe-ions of variable valency, their resistivity is greatly dependent on partial pressure of oxygen. In the reducing environment, variations in olivine resistivity produced by heating and cooling are in good agreement. This indicates the absence of thermochemical transformations which could have influenced the mechanism of conductivity in these minerals.

Proceeding from the ultrabasic composition of the upper mantle material, let us consider the relationship $\varrho = f(p, T)$ for dunites and peridotites. Fresh unserpentinized dunites and peridotites, owing to a high magnesium content, are characterized by a high resistivity and linear relationship of lg ϱ and $1/T$ throughout a temperature range of 100—1200 °C (Fig. 3). The presence of chromium cations in dunites increases their resistivity as compared to peridotites, because chromium cations do not participate in the Fe^{2+} and Fe^{3+} charge exchange.

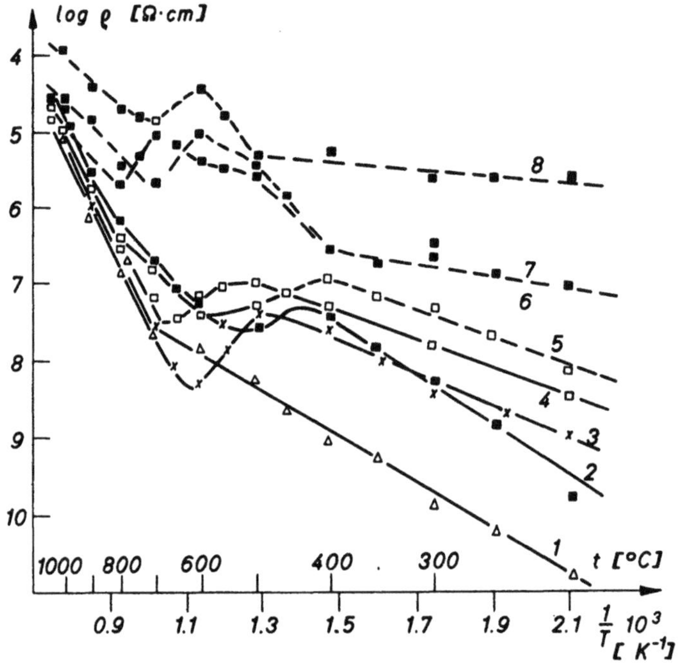

Fig. 3. Temperature dependence of ϱ, serpentinous ultrabasic rocks; *1—5* — dunites; *6—8* — peridodites.

Serpentinized ultrabasic rocks may have both high and low resistivity. Low resistivity is observed in the rocks whose serpentinization was accompanied by the emplacement of metallic minerals on grain peripheries or in the form of intersecting streaks which form current-conducting channels.

In the interval of 500—800 °C, serpentinized rocks have typical temperature regions of anomalous resistivity variation resulting from dehydration and chemical reactions involved due to the presence of free state iron oxides.

In the region of intrinsic conductivity the difference in the resistivity of fresh and serpentinized rocks is smaller. Peridotites, olivinites and pyroxenites usually exhibit an insignificant decrease in resistivity and activation energy under quasihydrostatic pressure of 1 to 20 kbar in a temperature range of 100—650 °C [148].

Xenoliths of ultrabasic rocks and eclogites are of particular interest in the study of the composition and physical properties of the upper mantle material. As evidenced by petrochemical analysis, xenoliths found in the kimberlite of the "Obnazhennaya" pipe are the representatives of pyropebearing rocks. A distinctive feature of these rocks is a high resistivity in a broad interval of temperatures (Fig. 4). This behaviour is determined by the content of about 50 per cent of magnesium-rich garnet (pyrope) and clinopyroxene with chromium. Owing to a large contribution of an electronic component, the resistivity of almandine eclogite and olivine xenoliths is lower, particularly, in a high temperature range of 800—1200 °C.

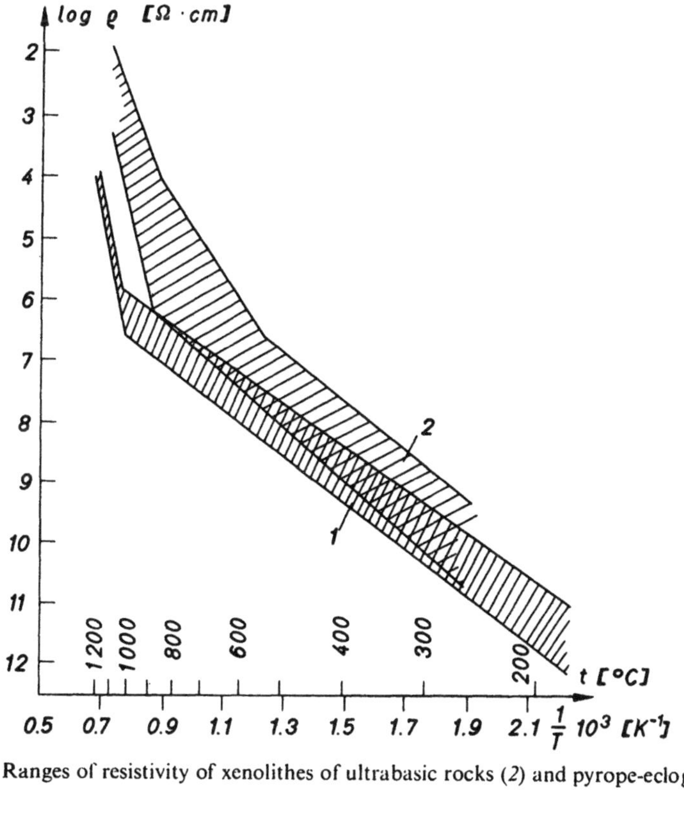

Fig. 4. Ranges of resistivity of xenolithes of ultrabasic rocks (*2*) and pyrope-eclogite (*1*).

Thus, the mantle material believed to be an association of olivine, garnet (pyrope) and pyroxene appears to have a large range of resistivity values varying at 800—1400 °C depending upon the presence of Fe-cations and a Fe^{2+}/Fe^{3+} ratio. The effect of pressure to 40 kbar upon the resistivity of the indicated mineral association is not large at these temperatures, and appears to be compensated in the Earth's interior by the increase of temperature with depth.

As mentioned above, the resistivity of the upper mantle material may be substantially affected by various physicochemical processes that are believed to occur in the Earth's interior. Most important among these are melting and amorphism of the material. Experimental data show that in the transition of the material from crystalline to amorphous state, its resistivity may decrease by one to five orders depending on the chemical composition of the material. In the experiments on basalts compositionally varying from high-alumina to alkaline, the greatest effect of melting on restivity variation was observed in high-alumina most resistive basalt (Fig. 5). In this case resistivity decreased by more than two orders [152]. Pressure virtually does not increase the resistivity of high-alumina crystalline tholeiite, but displaces the melting point by 70 °C. Whereas the effect of pressure on the resistivity of quartz tholeiite is rather strong, as seen in Fig. 5. These data are not contradictory but consistent with the results of magnetotelluric measurements which reveal a conductivity layer in the upper mantle at a depth of 60 to 300 km depending on the geological structure of a region surveyed.

Of no less importance to the understanding of the upper mantle geodynamics are experimental data on resistivity variation in the case of solid-phase transformations, such as polymorphism and chemical reactions.

Studying resistivity variation at a polymorphous transformation of cyanite-sillimanite (Al_2SiO_5) at pressure to 20 kbar, in [153] is recorded a change in resistivity at $t = 1130$ °C by not more than half order. And even this small variation was obtained as a result of extrapolation, while the experiment itself showed it to be smaller.

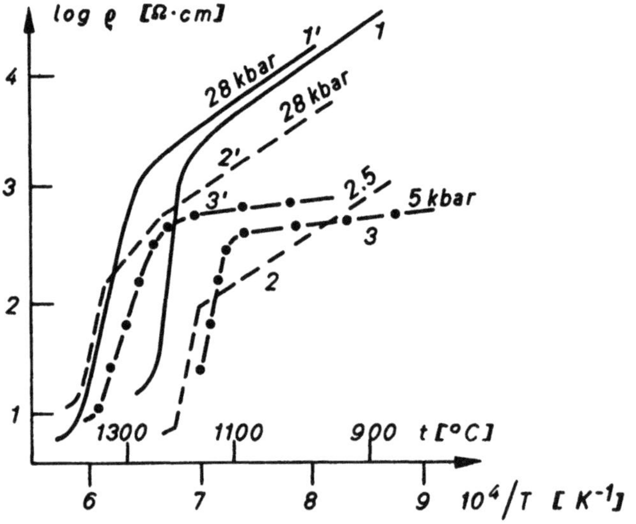

Fig. 5. Temperature dependence of basalts: *1, 1'*-high-alumina tholeiite; *2, 2'*-quartz tholeiite; *3, 3'*-alkaline tholeiite.

When heated, jadeite separates into albite and nepheline in accordance with the reaction

$$\underset{\text{(jadeite)}}{NaAlSi_2O_6} \rightleftarrows \underset{\text{(nepheline)}}{NaAlSiO_4} + \underset{\text{(albite)}}{NaAlSi_3O_4},$$

for which an equilibrium diagram is available. The resulting conductivity isobars at 7 and 12 kbar show an increase of resistivity at $t = 220$—$250°$ (at 7 kbar) and a flat slope of the line $\lg \varrho = f(1/T)$ (at 12 kbar). When P and T parameters of the indicated resistivity anomalies were plotted on a phase equilibrium diagram, they were seen to be in good agreement with the equilibrium line [154].

These two examples prove the general conclusion that no significant change of resistivity must occur in polymorphous transformations of a dielectric-dielectric type. On this basis we can assume that not great resistivity variation must occur in the transition of an olivine modification of forsterite into a spinel one.

Resistivity is expected to vary in jumps in the transition dielectric-semiconductor or semiconductor-semiconductor. An example of such behaviour is observed in the transition of olivine fayalite to spinel fayalite. This transition is marked by a resistivity decrease by more than one order at $t = 620$ °C and $P = 59.8$ kbar [155]. It is not improbable, however, that the indicated resistivity variation is not produced by structural transformations but results from chemical reaction involved by the presence of FeO, as it is observed in siderite in a process of dissociation [156].

Interesting results have been recently obtained on the transformation of some dielectrics to conductors under very high pressures. Such materials related to the problems of geophysics are quartz, corundum and diamond [157, 158, 159].

Extremely important among the above mentioned phase transitions are dehydration and hydration processes. Being widespread in the Earth's crust, these processes are likely to

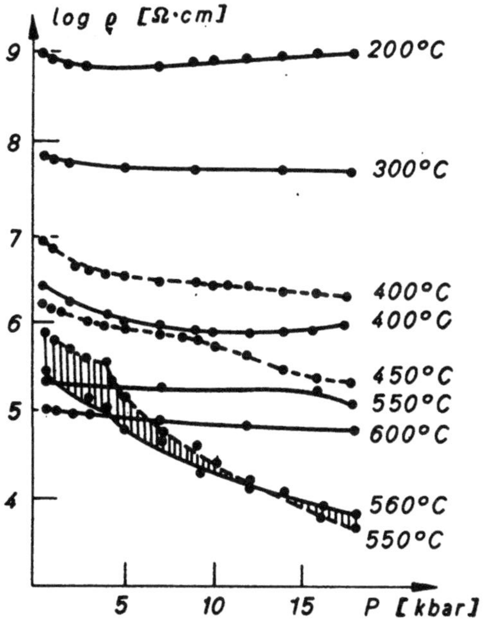

Fig. 6. Isotherme of ϱ of two samples of serpentinite under $P = 1$—20 kbar.

occur at the boundary of the crust and mantle, as manifested by the example of serpentinization of olivinites and dehydration of serpentinites. The latter process is accompanied by substantial changes in volume, strength, plasticity and resistivity, as seen from Fig. 6.

The observed intensive decrease of resistivity with temperature and also, under isothermal conditions, with pressure is, to a great extent, related to the presence of iron oxides. In the absence of the latter, as for instance, in brucite, no resistivity decrease is observed. In a closed system, the process of dehydrations is always accompanied by a decrease in resistivity at the expense of water present in the rock.

In conclusion, it is worth noting that resistivity is not only a sensitive indicator of temperature, but also of many physicochemical processes occurring in the upper mantle. Resistivity variation in jumps accompanies melting, exchange reaction between Fe^{2+} and Fe^{3+}, transition from dielectric to semiconductor and to metal; substantial resistivity variation is also caused by dehydration. This research is an example showing that correlation of theory, field results and experimental data is the only way to understand the processes going on in the Earth's interior inaccessible to direct observation.

LOW-FREQUENCY ELECTRIC SPECTROSCOPY OF THE SECOND-ORDER PHASE TRANSITIONS IN ROCKS

T. L. Chelidze, G. G. Geladze

It is common knowledge that such phase transitions as partial melting or polymorphous transformations (e. g., transitions from the rhombic to the cubic structure of spinel in olivine) may drastically affect the crust's and the upper mantle's structure causing creation on layers with specific physical properties. Therefore, the study of the electric results of such transitions seems to be very important in the physics of the Earth [160].

The method of electric spectroscopy used in this study consisted in investigation of the frequency dependence of dielectric permeability ε and electric resistivity ϱ. Measurements were made on samples, 0.5 × 2.5 × 2.5 cm, previously dried at 115 °C, using the bridge method, the frequency range and the temperature range being 0.7—200 kHz and 100—1000 °C, respectively. Established values of ε and ϱ were estimated, as the samples had been exposed to the given temperature for two or three hours, and then measured. The non-oxidizing contact on the samples was ensured by means of platinum electrodes applied by the cathode sputtering technique in vacuum. Both the ohmic properties of the contacts and the effect of the electrode capacitance and of other distorting factors [161] were kept under control.

The samples studied revealed electric effects in Curie point of magnetic and at the α—β-transition of quartz. In pure substances these transitions occur at T = 573 °C.

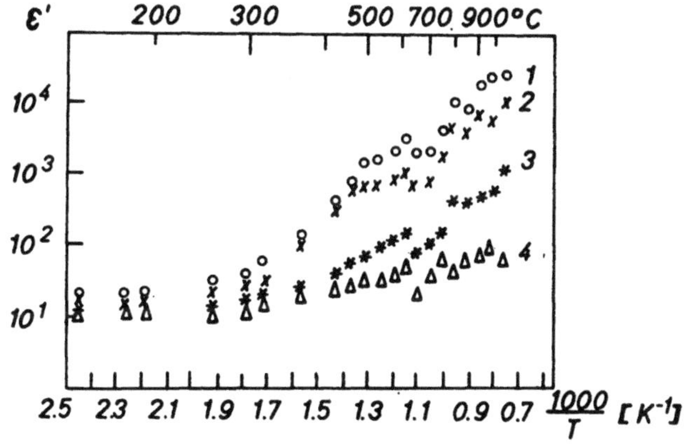

Fig. 1. Dielectric permeability of the garnet skarn (magnetite content ~16%). Frequencies: *1* — 0,7 kHz; *2* — 2 kHz; *3* — 20 kHz; *4* — 200 kHz.

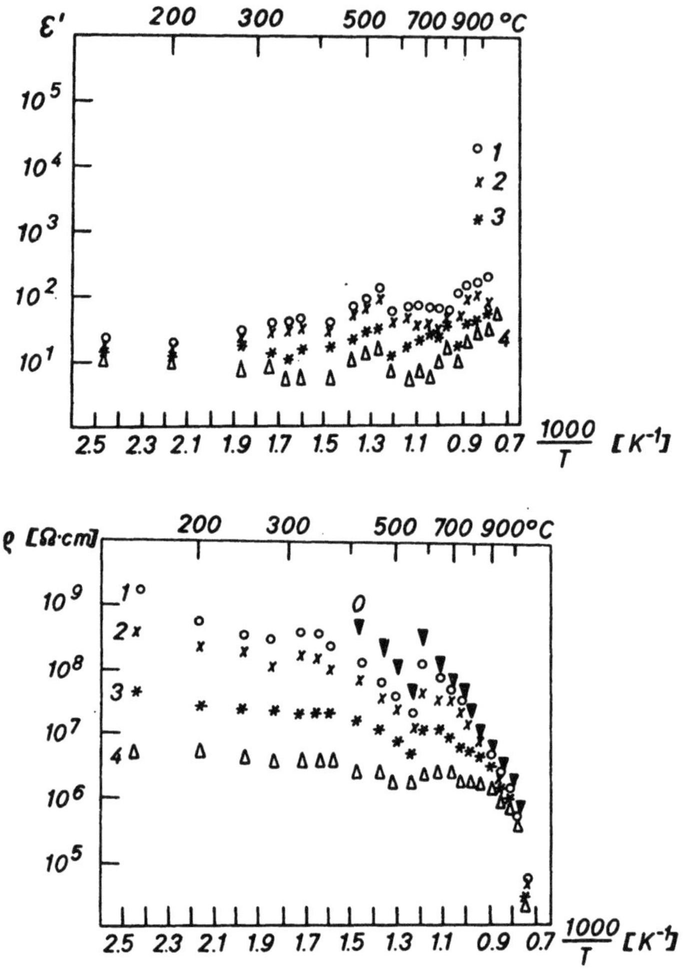

Fig. 2a. b. Electric effects for the α-β-transition of quartz. Granite No. 202. Frequencies: *0* — direct current, *1* — 0.7 kHz, *2* — 2 kHz, *3* — 20 kHz, *4* — 200 kHz.

Figs. 1—2 of results of the measurements show clearly the singularities existing along $\varepsilon(T)$ and $\varrho(T)$ in the region of the critical temperature T_c, as typical λ-curves. Differential thermal analysis has proved the existence of phase transitions of magnetite in sample 10 and of quartz in sample 202, right in this high temperature region. The slight deviation of T_c from 573 °C and the effect itself being greater than in pure quartz [162, 163] especially on the $\varrho(T)$-characteristic can be explained by concurrent phenomena, such as dehydratation, rapid diffusion of impurities.

Some authors explain the anomalies ("humps") in the temperature characteristics of electric conductivity of some methamorphic rocks as due to the dehydratation processes, i. e., the outflow of crystallization and constitution water from the minerals' lattice [164, 165].

The effects observed in this study on granite (Fig. 2) could hardly be assigned to dehydratation since: 1 — no singularities were found on $\varepsilon(T)$ and $\varrho(T)$ at 570 °C on

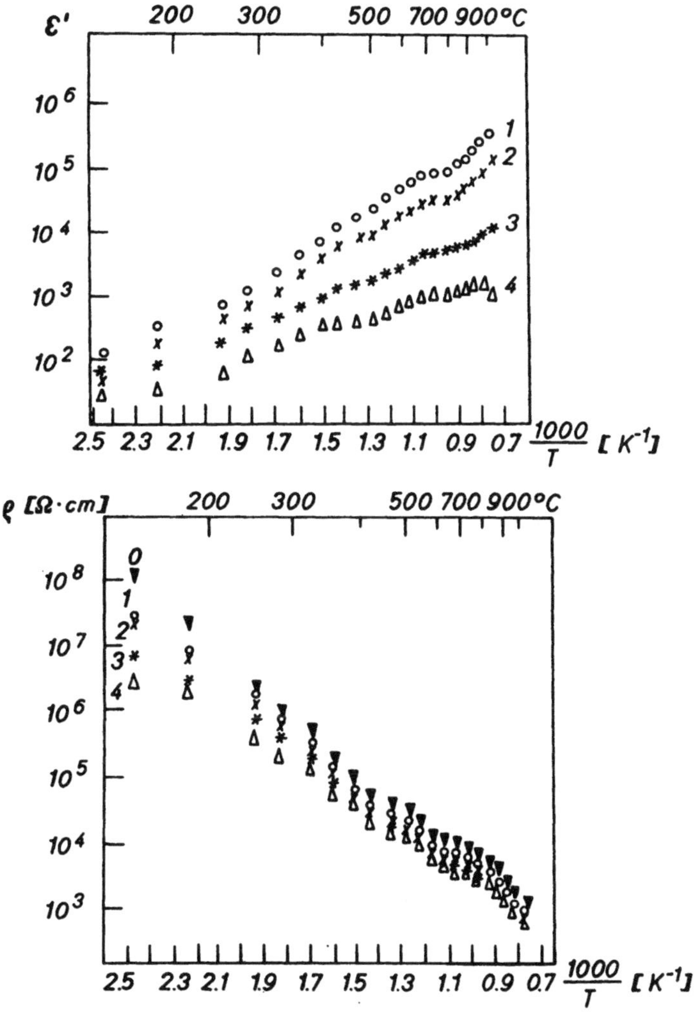

Fig. 3a, b. Temperature dependence of ε and ϱ. Granite No. 200. Frequencies: *0* — direct current, *1* — 0.7 kHz, *2* — 2 kHz, *3* — 20 kHz, *4* — 200 kHz.

sample 200 where thermograms exhibited large weight losses (Figs. 3, 4), while they were evident on samples with much lower dehydratation (sample 202); 2 — the $\varepsilon(T)$- and $\varrho(T)$-characteristics are shown below as approximated by a function describing phase transitions of a higher order. It is quite possible as well that dehydratation increases the α—β-transition effect.

Let us find an adequate mathematical model for the behaviour of ε and ϱ in the critical region. It is a modern concept that behaviour of such physical properties of a substance as magnetic susceptibility, compressibility, degree of ordering, etc., in the region of the critical temperature T_c, can be described by the following power function [166, 167]:

$$f(T) \sim (\Delta T)^{\lambda}, \quad \text{where} \quad \Delta T = T - T_c \tag{1}$$

and λ, the so-called critical index, can vary for $T < T_c$ and $T > T_c$.

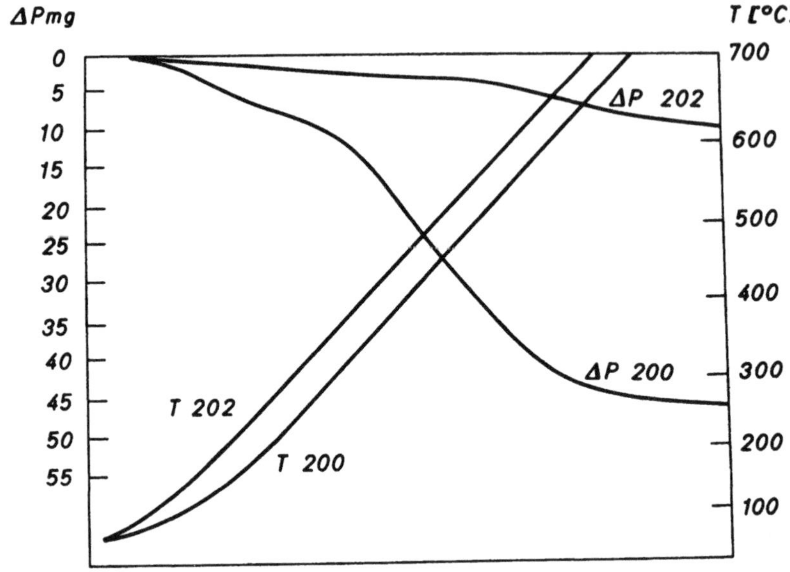

Fig. 4. Thermograms of samples 202 (1) and 200 (2), weight losses 0.91 % and 2.75 %, respectively.

This similarity in behaviour of various thermodynamic values in the neighbourhood of T_c implies the universal character of all the occuring phenomena, namely, an increase of fluctuations correlation radius. The reculiarity of the system is expressed in the difference between the T_c-values for specific systems [166]. Weiss' classical theory provides the critical index γ of the order of 1 with $T < T_c$ for the isothermal susceptibility. Having plotted the function $\lg \varepsilon \left(\lg \frac{T - T_c}{T_c} \right)$ for sample 10 where the phase transitions of magnetite were observed, we obtained the same value of γ as from the magnetic measurements [167], $\gamma \simeq 0.93$ (Fig. 5).

Straightening of the $\varepsilon \left(\frac{T - T_c}{T_c} \right)$ on the bilogarithmic scale for granite samples with the α—β-transition of quartz provided approximate values of the critical index γ for ε of the order of 0.6—0.7 (Fig. 6). The quality of the strainghtening is not good; probably the relation of ε to (ΔT) is not a pure power function, but rather like

$$\varepsilon(T) \sim (\Delta T)^{\lambda} \cdot f_0(T) , \tag{2}$$

where $F_0(T) = f_0 + f_1(\Delta T) + f_2(\Delta T)^2 + \ldots$ is analytical in the region of $T = T_c$.

Behaviour of such kinetic properties of a substance as the electric conductivity in the region of the transition point is not sufficiently well studied. Strong deviations of ϱ have been noticed in this study in the α—β-transition region of quartz for granite and gneiss, the amplitude of the " hump" reaching one or two orders. Plotting a relation for ϱ similar to (1) gives the critical index value 0.5, close to that for ε, indicating that the electric transfer is obviously estimated by the growth of the fluctuation correlation radius, T_c being approached from the lower temperatures. For $T > T_c$, the resistance seems to increase sharply, probably due to an increased scattering of the carriers during the transition from an ordered low-temperature phase to a disordered high-temperature phase. The

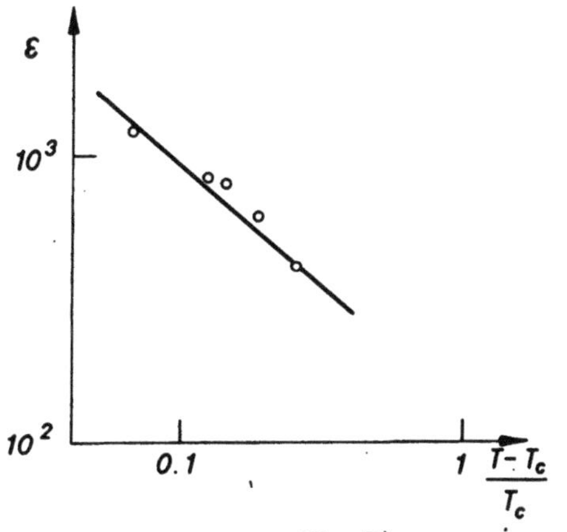

Fig. 5. Relation of lg ε to $\lg\left(\frac{T-T}{T_c}\right)$ for the garnet skarn.

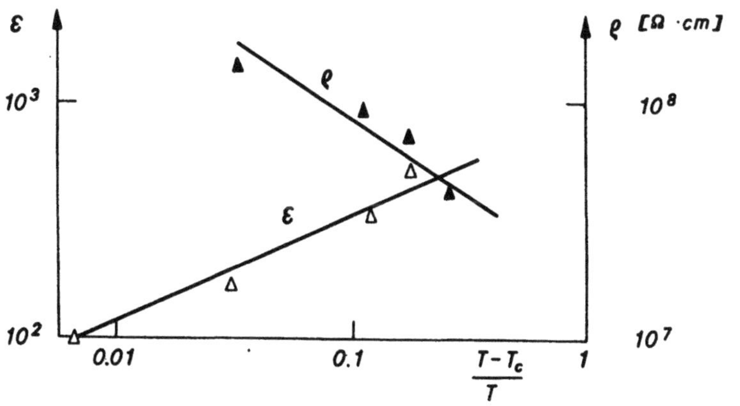

Fig. 6. Relation of lg ε and lg ϱ to $\lg\left(\frac{T-T}{T_c}\right)$ for granite No. 202.

electric transfer at $T < T_c$ is likely to be of the electronic nature, since this temperature region includes a strong dependence of activation energy E_a on f, this E_a reducing as low as the tenths of an electron volt at higher frequencies (Fig. 2).

The above transitions have been almost equally well expressed at all the frequencies used.

The following geophysical effects may be caused by the phase transitions studied in this work:

a) forming of zones of higher conductivity either in the crust or in the upper mantle due to the α—β-transition of quartz at about 550 °C isotherm depth (neglecting the change of T_c under pressure). In the Caucasus this isotherm passes at the depth of 25—30 kms where such rocks as granite are quite possible. Although the ε and ϱ peak is considerably sharp, even sometimes being a point peak (Fig. 2), the shift of the T_c-value in accordance with the pressure should be followed by widening of the α—β-transition region, consequently causing the creation of layer of a higher conductivity.

b) as far as at T_c the second derivatives of thermodynamic potentials over thermodynamic forces (the so-called stability coefficients) pass through the minimum [168], it is right in this area that the system detects maximum susceptibility to the external effects, i.e., low forces involve strong reactions. The same is true as well for the inverse value of dielectric permeability, $1/\varepsilon$, and magnetic permeability, and also for such basic features as compressibility which can be well described by (1), i.e., compressibility increases sharply at T_c. Since quartz constitutes a significant part of the granite-type rocks, the α—β-transition may involve drastic changes in the whole formation's compressibility. This effect can be directly related to the problem of the earthquake mechanism and may influence the seismic wave propagation. Therefore, unlike the first-order phase transitions affecting the strain in the interior of the Earth by changing the volume, the higher-order phase transitions affect the state of stress and velocities of the elastic waves due to the changes in compressibility.

SURFACE CONDUCTIVITY AND POLARIZATION OF ROCKS AT HIGH TEMPERATURES

T. L. CHELIDZE, G. G. GELADZE

The problem of partial melting in the crust's and upper mantle's rocks, and of the effect of this phenomenon on the physical properties of the Earth's material has become a matter of great interest in the recent years. The conducting astenospheric layers at the depth of about 100 km are considered as resulting from that partial (or should we call it "surface") melting [181, 182]. The surface liquification involves considerable geodynamic processes as well.

Electric conductivity γ and dielectric permeability ε are rather sensitive to the effect of the surface melting, therefore their study seems to be of an obvious importance. The effect of the conducting shells on the electric properties of disperse systems is well known [183—186], but it was not until work [187] was published, that a remarkable similarity of experimental frequency dependences of γ and ε of rocks was notices during such completely different processes as moistening and heating. This similarity can be well justified physically heating causes melting first of the grain surfaces due to the following factors: a) an excess surface energy of the grains; b) existance of vacancies facilitating migration of particles on the surface. The ratio of the surface ($\mathfrak{D}_s$) and bulk ($\mathfrak{D}_0$) diffusion coefficients at the temperature close to that of melting, T_0, is about 10^4 and increases with the reduction of T. Since in solids $\mathfrak{D}_0 \sim 10^{-9}$ cm^2/sec, with $T \to T_0$, the value of $\mathfrak{D}_s$ is 10^{-5} cm^2/sec, i.e., of the order of the diffusion coefficient in electrolyte.

So, the "surface melting" can be quite successfully simulated by moistening the system.

The effect of the surface shells is accounted by means of additional terms in the expressions of the theory of heterogeneous dielectrics (Maxwell-Wagner theory). For example, a spherical particle with the radius a and conductivity γ_2, covered with a shell of thickness h and conductance γ_l, can be replaced by an equivalent homogeneous particle with conductivity γ_2' [186]:

$$\gamma_2' = \gamma_2 + 2\frac{h}{a}\gamma_l = \gamma_2 + \frac{2K_s}{a}, \tag{1}$$

where γ and K may be complex in the alternating fields; $K_s = h\gamma_l$ is the surface conductance.

Such expressions enable an easy explanation of any deviations of the experimental results from the theoretical ones, in case the anomalies are connected with the surface contributions.

Following the analysis of the currents of substance charges, heat, etc., in the system, it is possible not only to make the above model but to calculate the relaxation time τ as well. For the conducting film on the insulator's grain τ is specified by two processes — diffusion and migration — of particles in the film [188]:

$$\tau = \frac{a^2}{2\mathfrak{D}_s[1 + \chi^2 ah\varepsilon_f/(\varepsilon_2 - 2\varepsilon_1)]}, \tag{2}$$

where $\mathfrak{D}_s$ is the film's diffusion coefficient, χ is the reverse Debye screening radius, ε_f, ε_2, and ε_1 are the dielectric permeabilities of the film substance, the grain and the continuous medium, respectively. The velocity of the process is defined by diffusions for $\chi a \ll 1$, and by film conductivity for $\chi a \gg 1$. For not too small particles (Fig. 1) $\chi a \gg 1$ and

$$\tau = \frac{a(\varepsilon_2 - \varepsilon_1)}{8\pi K_s}. \tag{3}$$

Fig. 1 shows the relation of $\varrho = 1/\gamma$ and ε of the disperse quartz to the grain size at $T = 1000$ °C and 2 and 200 kHz; ϱ is decreasing with the grain diameter d almost three times, while ε increases 15 times, with the volume fraction of quartz $\varphi = 0.6$ remaining stable. This proves the existence of surface contributions to conductivity and polarization. Having admitted that for the largest fraction ($\sim$1 mm) the surface transfer is negligible and the conductivity is precisely equal to the theoretical Maxwell-Wagnerian value (γ_{mw}), one can calculate the excess surface conductivity γ_s [189]:

$$\gamma_s = \gamma_2 - \gamma_{mw} \tag{4}$$

where γ_2 is the observed value of the electric conductivity.

Since for large fractions $\gamma_s = 0$, for the small ones $\gamma_s \simeq 3 \times 10^{-6}$ ohm/cm. Hence the surface conductivity can be calculated:

$$K_s = \gamma_s \cdot \varphi_g/S, \tag{5}$$

where $\varphi_g = (1 - \varphi)$ is the porosity which is equal here to 0.4. When S is of the order of 5×10^3 cm²/cm³, meaning small fractions (Fig. 1), we obtain $K_s \simeq 2 \times 10^{-10}$ mho. It should be noted that in an earlier study [185] we obtained $K_s = 10^{-9}$ mho for the same disperse quartz moistened by electrolyte forming a thin conducting surface film.

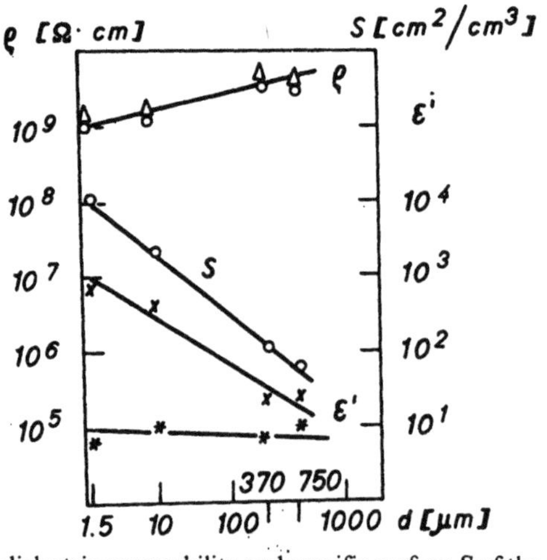

Fig. 1. Electric conductance, dielectric permeability and specific surface S of the disperse quartz in relation to the grain's size. Frequencies for ϱ: 0 — 2 kHz, Δ — 200 kHz; for ε: x — 2 kHz, * — 200 kHz.

Theoretical estimation of the relaxation time is possible by (3) when K_s is known; for the grains with $a = 10$ µm we get $\tau \simeq 3 \times 10^{-5}$ sec. Therefore, dispersion of ε is to be expected at low frequencies of the order of units and tens of kilohertz, and is to vanish at $(\omega\tau)^2 \gg 1$, i. e., at hundreds of kilohertz. Even though the theoretical relaxation time agrees well with the experimental one, the above film model is incapable of providing an explanation to the dielectrical increment value reaching thousands of CGSE at low frequencies. Meanwhile, maximum possible theoretical values of ε of disperse quartz are two orders lower according to this model and taking into consideration the screening effect of the conducting film:

$$\varepsilon_{max} = \varepsilon_1/(1 - \varphi)^3 \simeq 16 .$$

ε_1 being here the dielectric permeability of the continuous medium (air), $\varepsilon_1 = 1$ and ε of the grains admitted as infinitely high. Besides, it is difficult to explain the almost linear correlation of conductivity and dielectrical permeability in the film model.

It was not only the disperse quartz that reveals abnormally great low-frequency values of ε at high temperatures, and a symbasic increase of ε and conductivity, but all the other rocks studied, as well. This fact evidences the existence not only of film polarization but of some much stronger mechanism as well.

This mechanism may be the diffusions-migration polarization of the double electric layer of charged particles; these charges can be either connected with the surface (Schwartz-model) [190], or freely exchanging with the volume (Dukhin-Shilov model) [191]. The relaxation time in these models is expressed in a similar way:

$$\tau = a^2/\mathfrak{D} , \tag{6}$$

where a is the particle's size, $\mathfrak{D}$ is the surface diffusion coefficient for the first model and the bulk diffusion coefficient for the second one.

We have found that the double layer polarization model remains true with charged microheterogeneities of the specific size a on the surface of the particle [192]. At high temperatures the bulk diffusion coefficient $\mathfrak{D}_0 \simeq 10^{-9}$ cm²/sec. in solids, therefore within the used range of frequencies the heterogeneities with the specific size a from tens to hundreds of Ångström will relax (i.e., induce dispersion of ε); this might be dislocations, defect accumulations, crystallite boundaries, etc. [193].

The diffusion-migration model of polarization of microheterogeneities double layer can justify both the abnormally large value of the dielectric increment in rocks, and strong correlation of the conductivity and polarization [194]. In fact, following Maxwell-Wagner theory, the increment of the complex dielectric permeability $\Delta\varepsilon^*$ of a heterogeneous system is

$$\Delta\varepsilon^* = 4\pi N\varepsilon_1^* d^*/E_0 \tag{7}$$

where N is the efficient dipoles number in a unit volume, $\varepsilon_1^* = \varepsilon_1' - i\varepsilon_1''$ is the complex dielectric permeability of the continious medium, d^* is the complex dipole moment, E_0 is the average microscopic field.

For the real part of $\Delta\varepsilon^*$, that is for dielectric increment we have

$$Re\, \Delta\varepsilon^* = 4\pi N(\varepsilon_1' \cdot Re\, d^*/E_0 - \varepsilon_1''\, Im\, d^*/E_0) . \tag{8}$$

Since loss factor $\varepsilon_1'' = 4\pi\gamma/\omega$, then for the frequency $\omega \to 0$ the value ε_1'' goes into infinity, if the medium is a conducting one. Then, if only $Im\, d^*/E_0$ is not negligibly small, i.e., if the polarization field of a particle shows a considerable phase lag behind the applied field, for $\omega \to 0$ the second term in Eq. (8) may be very large, thus accounting for the large value of ε at low frequencies. Therefore, find that the medium's conductivity strenthens the weak polarization by means of the induced currents lagging in phase behind the primary field, thus accounting for the simultaneous growth of conductivity and dielectric permeability of the system.

Experimental test on the models of the spherical charged particles suspension and of the moisture-saturated disperse quartz gave a good agreement with the above theoretical concepts [195, 196].

It follows from the above discussion that the surface phenomena on the boundaries of grains and microheterogeneities (dislocations, accumulations of defects, etc.) can specify the electric properties of the Earth's substance at high temperatures. These phenomena seem to us very important due to other significant physical properties, such as the rheologic, thermal and other characteristics, since these properties are specified by migration and diffusion just as the electrical properties.

MSS STRUCTURES IN THE Fe-Ni-S SYSTEM AND THEIR ELECTRICAL CONDUCTIVITY AT HIGH PRESSURES AND TEMPERATURES

A. KRAFT and U. SEIPOLD

Introduction

The investigations of the Fe-Ni-S ternary system are moved on by two special disciplines within the geosciences. The iron nickel sulphides are an important group of minerals, because many representatives of this group occur as accessory minerals in a great number of rocks, while others again are main constituents of large ore deposits. Here the monosulphides are of special importance, but at the same time are also most contradictory in their structure and stability. Accordingly there is extensive special literature, the bulk of which however deals with the end representatives of the monosulphides $Fe_{1-x}S$ and $Ni_{1-x}S$ [197—200]. Investigations of the corresponding ternary compounds in most cases deal with the stability of the economically important pentlandite phase [201, 202], which at low temperatures occur together with the monosulphide solid solution.

During the past few years this ternary system has found a great interest as a result of recent theories and experimental results concerning the theory of the earth's core. Here the phase stability, the melting behaviour as well as the conductivity at high pressure are of great interest.

Phase investigations in MSS samples

The phase diagram of the Fe-Ni-S ternary system

In the region of monosulphides the phase diagram is highly temperature-dependent. Fig. 1 shows a cross-section at 600 °C according to [203]. A characteristic feature lies in the band of the monosulphide solid solution (MSS), which extends over the diagram from 992 °C down to less than 300 °C. At 300 °C it is still completely miscible, but its width is only one third of that at 600 °C. Till now miscibility limits were not observed.

Below 400 °C the two end members $Fe_{1-x}S$ and $Ni_{1-x}S$ are subject to phase transitions. For $Fe_{1-x}S$ it ist the magnetic β-transition at 315 °C and the structural α-transition at 137 °C into the 3 A; 2 C superlattice structure.

The MSS and their two end members crystallize in the B-8 type (NiAS). Especially in the case of $Fe_{1-x}S$ a number of superlattice structures occur which shall be dealt with below. Although the occurrence of superlattice structures in the MSS is a confirmed fact, it was not yet subject to a systematic investigation [201]. $Ni_{1-x}S$ exists only in two known forms, the high-temperature of β-form, which is isostructural to pyrrhotite, and the stoichiometric rhombohedral millerite structure. At temperatures below 282 °C the high-temperature form

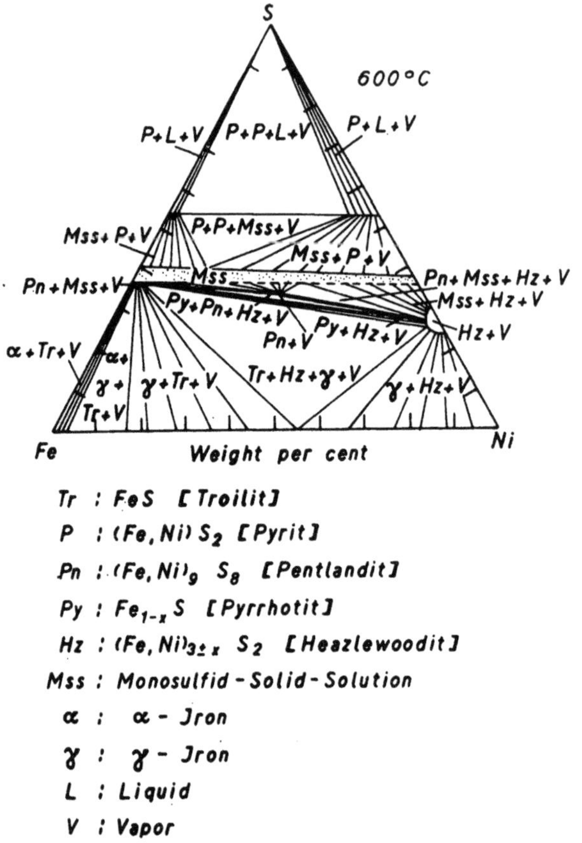

Fig. 1. 600 °C isothermal section of the condensed Fe-Ni-S-system.

is eutectically transformed into NiS and Ni_3S_4 [197]. Both end members as well as the MSS occur in nonstoichiometric forms, $Fe_{1-x}S$ having the largest range of variation.

Experimental work

Synthetic samples were exclusively used for the experiments. The samples were molten down at 600 °C in evacuated quartz tubes. The starting materials used were either the pure elements or an $Fe_{1-x}S$-$Ni_{1-x}S$ mixture. The results presented here exclusively refer to samples which were synthetized from the elements. After a reaction time of 24 hours they were opened, pulverized and after-annealed over 1 week (Set 1 — MSS — 1 → MSS — 8). Then the phase determination was carried out by X-ray-techniques with a counter diffractometer in step-scanning operation.

The B-8 structure (NiAs type) with the hexagonal lattice constants a ≈ 3.44 Å; c ≈ 5.88 Å is the starting point for the structural phase determination. All other occuring strutures of the MSS can be reduced to this basic structure and are treated as superlattice structures mainly by varying the *c*-parameter.

A different multiphase condition is observed only in the central portion of the MSS band, being due to the occurrence of the pentlandite phase (Fe, Ni)$_9$S$_8$, which is cubically crystallized and isostructural to Co$_9$S$_8$.

The thermal stability of this phase was investigated in [204] by means of quenching tests and differential thermal analysis as well as high-temperature X-ray diffraction. Syn-

Table 1. X-ray data of the pulverized sample MSS-8 (7.4% Fe; 57.1% Ni; 35.5% S)

No.	d [Å]	J [%]	phase
1	4.085	5	$Ni_{3-x}S_2$
2	2.976	90	pyrrhotite/MSS
3	2.697	15	pyrrhotite
4	2.6025	75	MSS
5	2.514	35	MSS*)
6	2.410	30	MSS*)
7	2.087	60	pyrrhotite
8	1.996	100	MSS
9	1.8315	5	$Ni_{3-x}S_2$
10	1.7175	45	pyrrhotite/MSS

*) superlattice structure lines

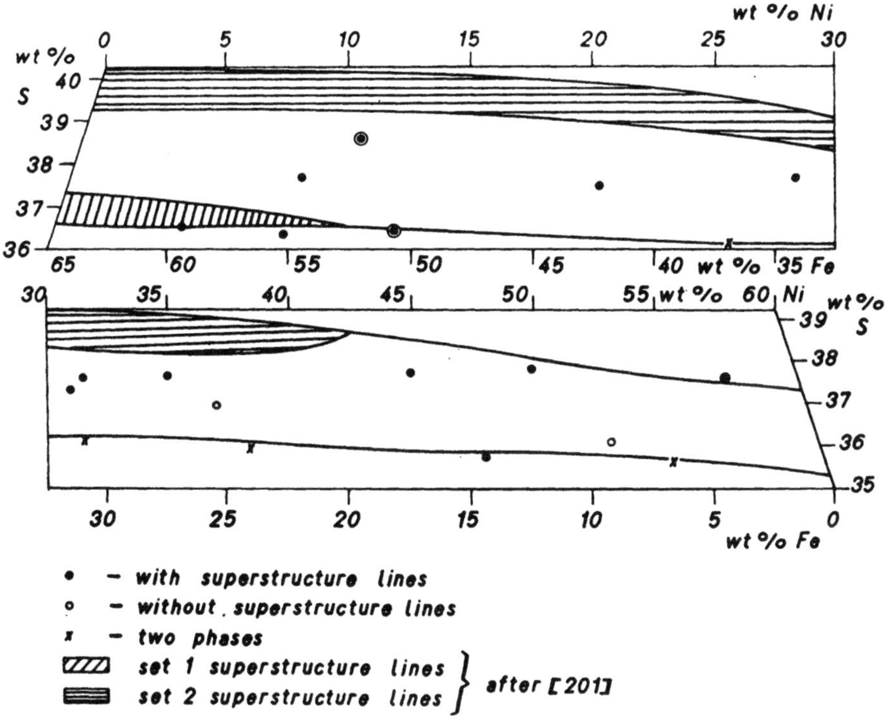

Fig. 2. MSS-Band across the center of the Fe-Ni-S-system.

thetic pentlandite of the composition $Fe_{4.5}Ni_{4.5}S$ at temperatures above 610 ± 2 °C segregates into $(Fe, Ni)_{1-x}S$ and $Ni_{3-x}S_2$ (high-temperature phase). This thermal limit decreases to 425 °C as the pressure increases to 36 kbar.

Pentlandite occurred as a second phase at the lower edge of the MSS-band of 25—38 wt % Ni, which corresponds to an Ni/Fe ratio of 0.71—1.58 in the samples synthetized here. This two-phase condition was observed in the samplės MSS-4, MSS-5, MSS-6.

In [205] is found a limit of 0.72—1.38, and in [206] is found a limit of 0.34—2.46 (Fig. 2).

Another multi-phase sample occurred at 7.4 wt % Fe (MSS-8) at the lower boundary of the MSS band. Here one has a pyrrhotite/$Ni_{3-x}S$/MSS mixture. The complete pattern is shown in Table 1.

Certainly the occurring pyrrhotite is not responsible for the two superlattice structure lines being encountered, because the intensities of these lines are too high, and hence the lines must be associated to the MSS structure.

When considering the superlattice structures, a great number of which occur in the MSS, as shown in Fig. 2, no systematic consequences as stated in [201] are observed.

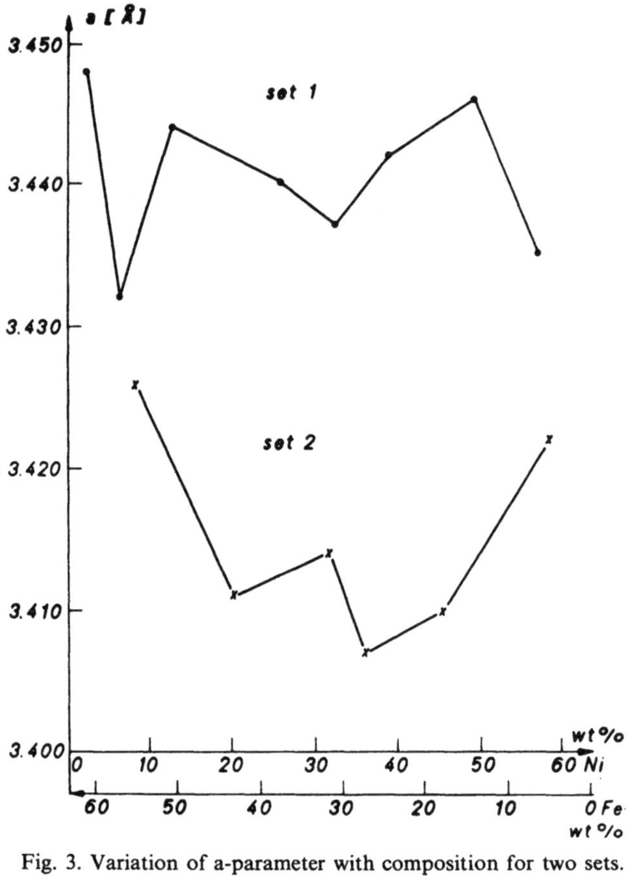

Fig. 3. Variation of a-parameter with composition for two sets.

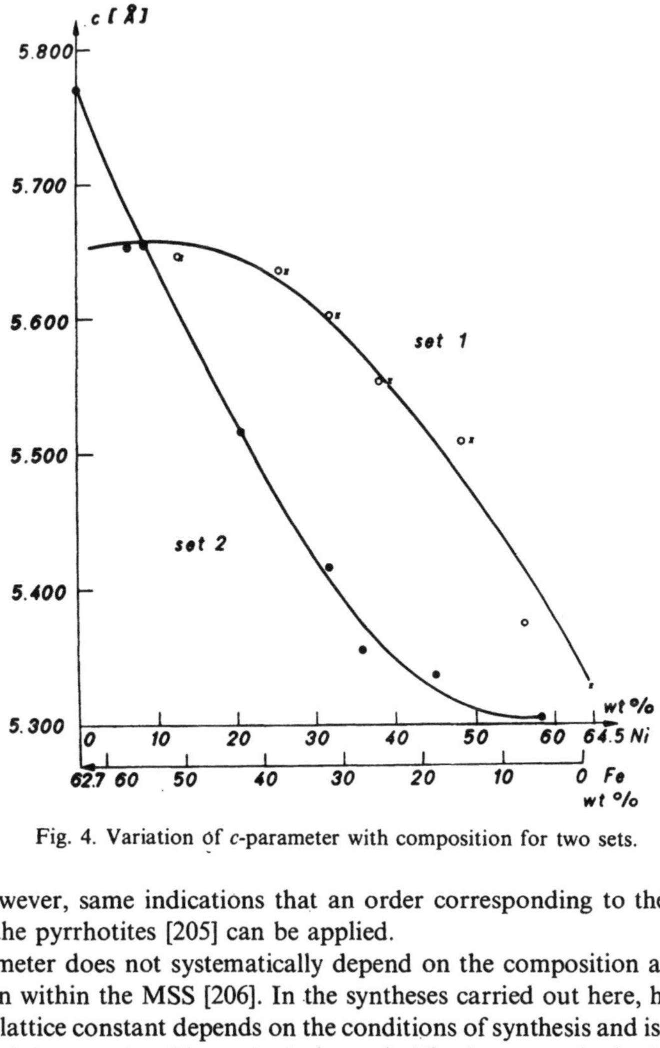

Fig. 4. Variation of *c*-parameter with composition for two sets.

There are, however, same indications that an order corresponding to the nC-superlattice structures of the pyrrhotites [205] can be applied.

The a-parameter does not systematically depend on the composition and shows only a slight variation within the MSS [206]. In the syntheses carried out here, however, it urned out that the *a*-lattice constant depends on the conditions of synthesis and is a measure of the homogeneity of the samples. The series 1 shown in Fig. 3 was synthetized according to the conditions described in the upper section. The series 2 was synthetized with the same duration, but without pulverizing and re-melting the sample after the primary reaction, which factor certainly influences the homogeneity. The *c*-parameter shows a clear dependence on the composition. However, in this case as well, different results were obtained for the series 1 and 2 (Fig. 4).

These results are of importance also for the investigation of the superlattice structures, because here it must be distinguished between stoichiometric structures, i. e., structures with an ordered distribution of interstices, and incompletely ordered superlattice structures. In [205] five stoichiometric structures are established for the pyrrhotites which are designated as superlattice structures associated to the B-8 structure with 4 *C* (Fe_7S_8); 5 *C* (Fe_9S_{10}); 11 *C* ($Fe_{10}S_{11}$); 6 *C* ($Fe_{11}S_{12}$); and 2 *C* (FeS). Add to this the 3 *C* structure, which is stated in [207] and derived from the B-8 structure with $a = 2A$; $c = 3C$. Here also the other superlattice structures were detected with a broader range of compositions.

While for $Fe_{1-x}S$ the problem of superlattice structures can be considered only under the aspect of stoichiometry, the situation is different in the case of MSS. Here two factors may be responsible for the formation of superlattice structures.

1. The purely stoichiometric conditions according to the general formula $Me_{1-x}S$ [Me $\triangleq$ (Fe, Ni)], which can effect an ordered vacancy configuration analogous to that considered for $Fe_{1-x}S$.
2. The type of diadochic incorporation in the formation of solid solutions, which may lead to a better than statistically ordered substitution and enables superlattice structure lines to occur.

However, the X-ray determination of the metal/sulphur ratio presents problems in the case of ternary compounds. The determination of non-stoichiometry, i. e. of the metal portion in monosulphides, by means of certain *d*-value dependence is well known for the binary compunds. Especially for $Fe_{1-x}S$, according to [208] the d_{102}-method has proven successful.

In [202], a dependence of d_{102} on the Ni concentration is given for the MSS. Since the *d*-values depend on both the metal/sulphur ratio and, as desribed here, on the iron/sulphur ration, and the conditions of synthesis likewise are of great importance for the lattice parameters, an exact determination of composition, as given by the d_{102} method for $Fe_{1-x}S$, does not appear probable. It seems to us that the c/a ratio provides a better dependence, but in this case the available measuring data and reference determinations are not yet sufficient.

Pressure and temperature dependence of the electric conductivity

The measurements of electric conductivity were carried out by means of a Bridgman opposed-anvil arrangement. An external furnace was used for annealing the samples. The temperature was measured in the immediate neighbourhood of the sample by means of an shielded thermocouple. The synthetized starting material was pulverized for the investigation. The powder was pressed into pellets having a thickness of 0.4 mm a diameter of 5 mm. To prevent the lateral extrusion during the measurement, the pellets were inserted into pyrophyllite rings having a diameter of 12 mm. This also results in a favourable pressure distribution within the sample. The deformation of the samples during the test is, however, still largely reversible. After the whole test had been completed, the sample thicknesses were still in the range from 0.36 to 0.39 mm. The measurements were carried out at constant pressure with a variation of temperature.

Since in most cases the sample resistivity was within the milliohm range, a Thomson bridge was used for the measurements. The highly ferriferous samples had such a high conductivity, that problems due to the contact resistances occurred. Therefore, a sandwich arrangement was used in order to obtain more favourable sample dimensions (length of 1 mm, diameter of 2 mm).

According to the result of the measurements of electric conductivity, the investigated compounds can immediately be subdivided into two groups by considering the absolute value of conductivity. The first group includes the samples MSS-1 to MSS-3, the second, the samples MSS-4 to MSS-8. The groups differ in that the resistivity of the latter groups exceed those of the former by 2 or 3 orders of magnitude. A detailed description shall be given below:

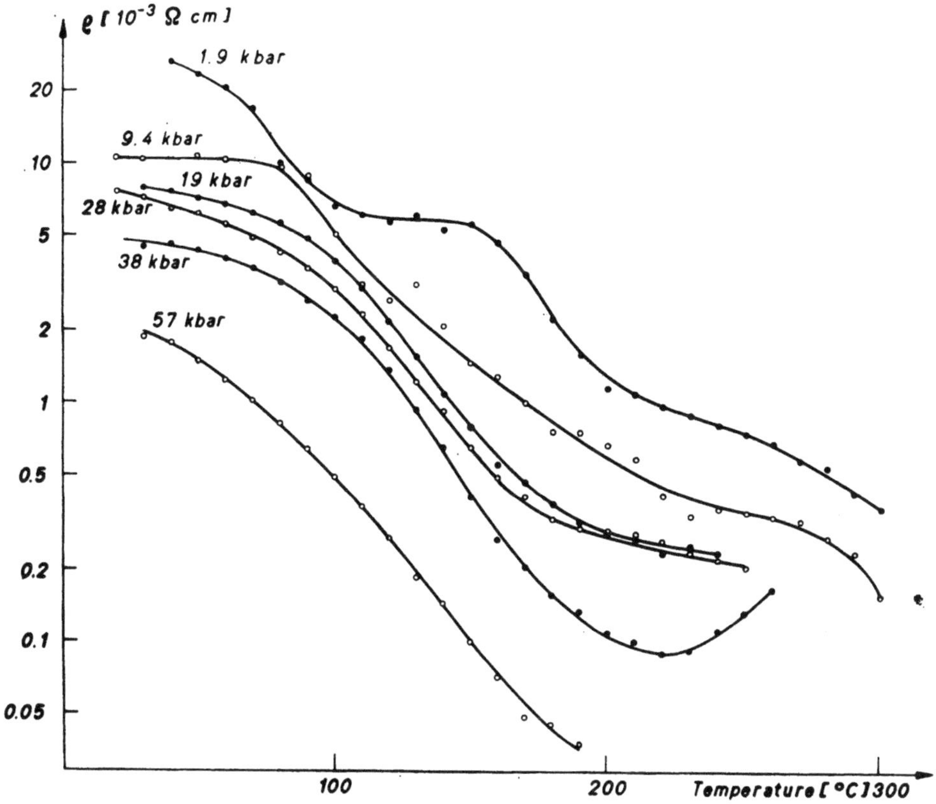

Fig. 5. Pressure and temperature dependence of electrical resistivity from MSS-1.

The electrical properties of the sample MSS-1 (Fig. 5) are critically determined by the occurrence of an anomaly, which is known as the α-transition in the case of FeS. In this case the resistivity is changed by 2 orders of magnitude. As compared to the behaviour in the case of FeS single crystals, the transition is distributed over a wider temperature interval, probably because of the different behaviour of the various crystallites in the employed polycrystalline sample. A certain broadening should also be caused by the inhomogeneous pressure distribution in the sample. As the pressure increases the transition is shifted towards lower temperatures. At the same time the curves are shifted towards lower resistivity as the pressure increases. For the anomaly, the absolute value of resistivity variation remains approximately constant.

At the low pressure levels (1.9; 9.4; and 19 kbar) the sample MSS-2 shows a qualitatively semiconductive behaviour (see Fig. 6), viz., its resistance decreases with increasing temperature. In the investigated temperature interval the resistivity changes by about a factor of 2. The α-anomaly is completely lacking. At higher pressure values the curves are shifted towards lower resistivity.

One has the unusual fact that the curves for the higher pressure values (28, 38, and 57 kbar) form a family, the resistivity of which exceed those of the first group. Moreover, the temperature dependence at the higher pressures is very low expecially at 38 and 57 kbar. The results suggest that essential conversions occur in the samples between

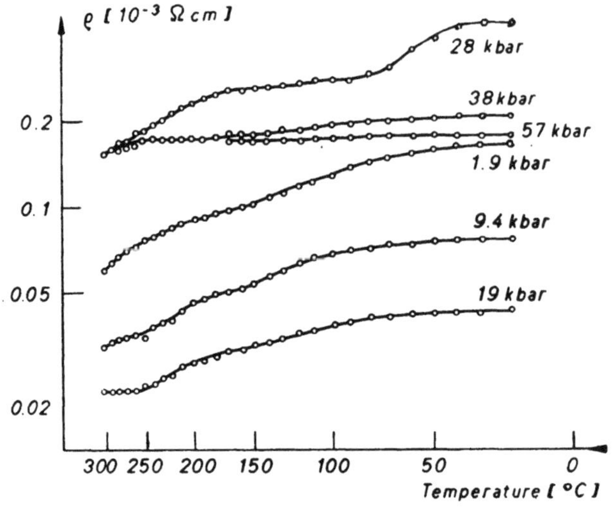

Fig. 6. Pressure and temperature dependence of electrical resistivity from MSS-2.

19 and 28 kbar. In the sample MSS-3, a semiconductive behaviour is observed at the low pressure values (1.9 and 9.4 kbar). However, the temperature dependence decreases with increasing pressure, so that at 28, 38, and 57 kbar the resistivity is nearly independent of temperature. The pressure dependence is normal, at higher pressure the curves being shifted towards lower resistivity.

As regards the electrical properties, the samples MSS-4, MSS-5, and MSS-6 show a relatively consistent behaviour. As an example, the results for MSS-4 are shown Fig. 7. A typical semiconductive behaviour is observed within large temperature ranges. The

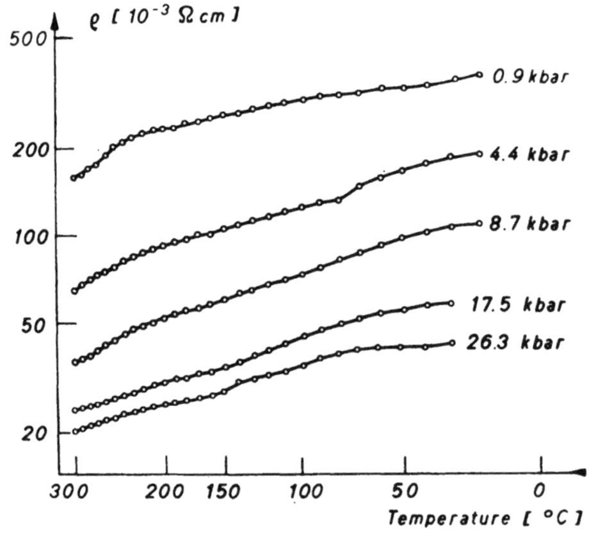

Fig. 7. Pressure and temperature dependence of electrical resistivity from MSS-4.

temperature dependence of the resistivity can be described by means of an exponential law:

$$\varrho = \varrho_0 \exp(-E_0/kT) .$$

The factor of the exponential function, ϱ_0, and the activation energy, E_0, are pressure-dependent. According to the qualitative predictions of the energy-band model, in most cases the activation energy decreases with increasing pressure. The pressure dependence of the factor of the exponential function is even more distinctly marked. The numerical values of both quantities are summarized in table 2 for the various samples for the respective pressure values.

At the high pressure the samples MSS-7 and MSS-8 are very similar to the group MSS-4 to MSS-6 with respect to their electrical properties. They are semiconductive, too. The corresponding parameters E_0 and ϱ_0 have been included in the table. A difference is found to exist at the low pressure values. The corresponding curves show relatively high resistivity values (see Fig. 8).

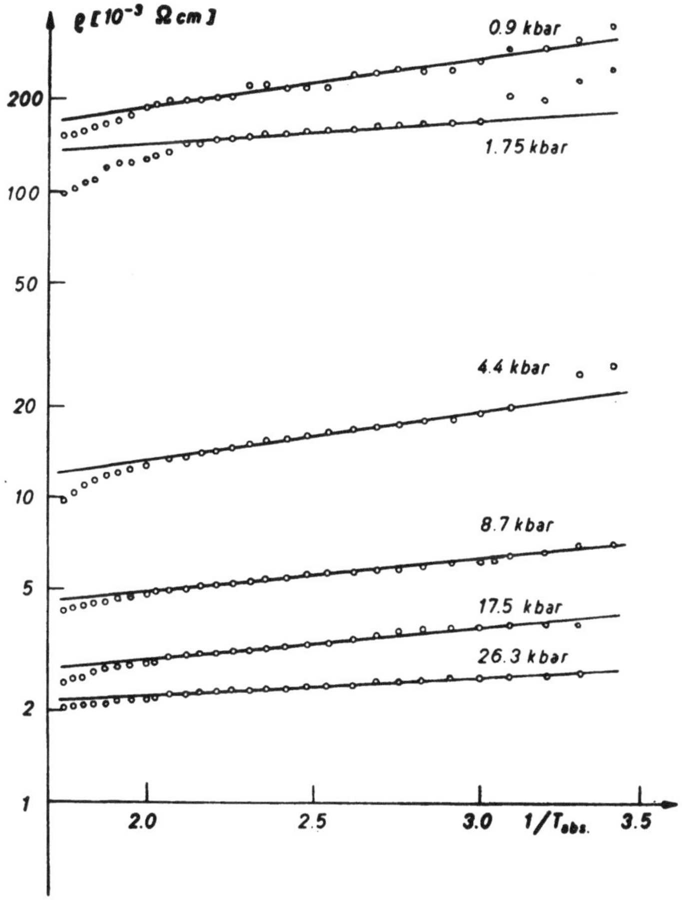

Fig. 8. Pressure and temperature dependence of electrical resistivity from MSS-7 und MSS-8.

Table 2. Preexponential Factor ϱ_0 and Activation Energy E_0

MSS-2			MSS-4			MSS-5		
p/kbar	E_0/eV	$\varrho_0/10^{-3}$ cm	p/kbar	$\varrho_0/10^{-3}$ cm	E_0/eV	p/kbar	$\varrho_0/10^{-3}$ cm	E_0/eV
1.9	0.11	0.0330	0.9	113	0.075	0.9	558	0.079
9.4	0.15	0.0112	4.4	26.5	0.124	1.8	694	0.048
19	0.08	0.0144	9	16.3	0.115	4.4	385	0.059
			17.5	10.8	0.107	9	233	0.048
			26	10	0.095	17.5	183	0.040
						26	133	0.041

MSS-6			MSS-7			MSS-8		
p/kbar	$\varrho_0/10^{-3}$ cm	E_0/eV	p/kbar	$\varrho_0/10^{-3}$ cm	E_0/eV	p/kbar	$\varrho_0/10^{-3}$ cm	E_0/eV
4.4	32.5	0.075	4.4	33.4	0.057	0.9	1170	0.061
9	17.5	0.112	9	20.8	0.048	1.8	1080	0.030
17.5	12.8	0.112	17.5	13.8	0.041	4.4	73.4	0.064
26	8.59	0.116	26	14.4	0.013	9	32.5	0.045
						17.5	20.0	0.042
						26	18.1	0.024

Discussion

In their electrical behaviour, the MSS of the iron nickel sulphides show a stronger dependence on the phases contained in them as compared to the composition dependence within the homogeneous MSS. The occurrence of relatively high absolute values is characteristic of the low-Ni portion (MSS-1—3). The existence of the α-transition in MSS-1 indicates a stoichiometric sample, because for FeS it was observed only within the limits FeS-$FeS_{1.04}$. This is supported by the structural data, because here only superlattice structure lines of the 2C type (troilite) occurred. This also supports the view that up to 10% Ni no substantial variations of properties occur in the pyrrhotites and MSS.

The electrical behaviour of the samples MSS-4, 5, 6 is determined by the occurrence of the cubic pentlandite phase, which makes an essential contribution to the observed semiconductor characteristic.

Superlattice structure lines did not occur in these samples. The sample MSS-8, the diffraction pattern of which was already described above, was of special interest. It shows several particularities. In spite of the extremely low Fe content, an NiS characteristic is not observed in the electrical behaviour. NiS has an increasing or nearly constant resistivity with increasing temperature, the pressure dependence being low. Add to this the high values of the factor of the exponential function at the low pressure levels (see Table 2) and the relatively high resistivity in the case of MSS-8. This suggests that a change of the multiphase equilibrium in the sample occurs. For clarifying this aspect, a squeezer photograph was taken at normal pressure and at 80 kbar, which is shown in Fig. 9.

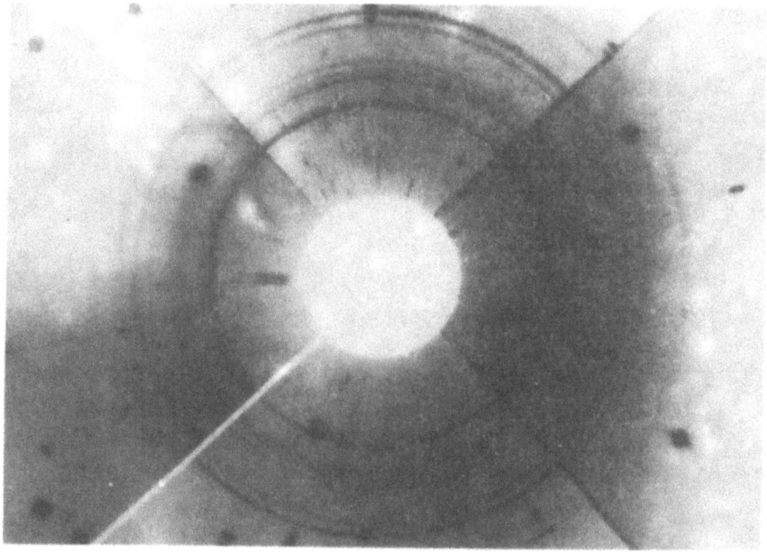

Fig. 9. X-Ray pattern from MSS-8 at normal temperature and 80 kbar.

Opposite quadrats of the film belong to the same photograph. Also, in this case a shift of the phase contributions in favour of pyrrhotite is observed, which is indicated by shifts of intensity in the line pattern. This suggests that pyrrhotite is the phase which is preferred with increasing pressure and that the MSS segregate to pyrrhotite under the action of pressure and temperature. Further investigations are expected to clarify this aspect.

CHAPTER IV

Magnetic and Thermal Properties

CHANGES IN MAGNETIC PARAMETERS OF IRON CHERTS FROM VARIOUS METAMORPHIC FACIES UNDER HIGH *P T*-CONDITIONS

T. S. Lebedev, B. Ya. Savenko, N. F. Poznanskaya and Yu. M. Stebnovskaya

Magnetic properties of iron cherts have been studied by many authors. Considered are mainly such problems as the distribution of magnetic characteristics of the said rocks in folded structures, correlation of magnetization with composition and texture of these rocks, nature of their magnetization application of magnetic parameters to interpretation of magnetic anomalies and so on [209, 213—215, etc.]. Only few publications [210—212, 222, etc.] contain the data about the effect of tectonic stresses on magnetic properties of iron cherts. Till recently, the variation of magnetic parameters of iron cherts under high pressure (P) and temperature (T) has been poorly studied. This information is important though in determining features of the thermodynamic regime under which iron chert has been formed and studying the state of magnetization in deep crustal zones.

The main aim of our investigations the results of which are presented below is studying the behaviour of magnetic parameters in iron cherts of different facies under the effect of high P and T.

Mineralogical studies

According to the existent notions, the physico-chemical conditions in which magnetite and iron chert of the Ukrainian Shield are formed depend on the sedimentation environment and intensity of the processes of regional metamorphism [210, 224]. Investigated were typical samples of iron chert formations metamorphosed to the green schist, amphybolite and granulite facies of the Great Krivoy Rog, Orekhovo-Pavlogradsky zone and Volodarsky magnetic anomalies. Iron chert of the green schist facies was sampled at a shallow depth in the Saksagansky region of the Krivoy Rog Basin; that of the amphibolite facies — from the Ingulets, Orekhovo-Pavlogradsky regions and deep-seated zones of the Saksagansky region. The iron chert of the granulite facies was sampled from the areas of the Pravobereghny and Volodarsky magnetic anomalies. Besides quartz and magnetite, typical rock-forming minerals in the iron chert of the green schist facies are carbonate, chlorite and hematite. For the iron chert of the amphibole facies the typical constituent minerals are amphibole and garnet, for that of the granulite facies is pyroxene. Each paragenetic association is characteristic of a certain geochemical degree with peculiar physical parameters (pressure, temperature and so on). Under condition of varying parameters of the medium where mineralization occurs, the same mineral (magnetite) appears in various generations differing in shope, structure and character of their distribution within the rock. Certain features of magnetite are observed as well in individual intercalations of iron chert. For this reason the samples to be studied were preliminarily out by layers (leaf-by-leaf) to pick up more pure and homogeneous variations of magnetite.

Mineralogical studies of the constituent magnetite in ferruginous formations of various facies have shown that, with growing degree of metamorphism, the size, shape and internal structure of this mineral change. In low-temperature associations of the green schist facies, magnetite occurs in the form of irregular isometric or angular grains of a size of 0.01—0.4 mm forming polyhedral intergrowths within the intercalations of ore. As a rule, the magnetite grains contain a large amount non-metalliferous inclusions (quartz, chlorite, carbonate) which often form a cribrate structure. The volume of the inclusions ranges from 10 to 50% and is directly dependent on size of magnetite grains. The noted features of the internal structure of the latter are also characteristic of silicate intercalations in ferruginous rocks which have undergone metasomatic changes.

In medium-temperature formations of the amphibolite facies, magnetite grains show irregular or idiomorphic outlines, the grain size ranging from 0.05 to 0.5 mm. Included in magnetite grains are usually amphibole and biotite. The inclusions are often regularly arranged, their volume is no more than 10—15% of the grain volume.

In high temperature rocks of the granulite facies, magnetite is associated with pyroxene and it almost lacks visible inclusions. Magnetic grains (0.5—1.4 mm) are observed as idiomorphic inclusions in pyroxene and form elongated lamellar secretions in the intergranular space. Certain lamellar magnetite grains under magnification by 1200 exhibit dotted inclusions of hematite.

The spectrum analysis of magnetite monofractions reveals a rather motley composition of impurities which amount to no more than tenths of a per cent. For magnetite of the low-temperature associations the following mineral impurities are characteristic: Al, Mg, Ca, Mn, Ni, Ti, V, Cr, Zr, Cu, Pb, Zn, Ge. In the high temperature formations, magnetite contains also Co, Mo, Sn.

The results of the nuclear gamma-resonance (NGR) studies have shown that magnetite of the low temperature associations lacks isomorphic impurities. The elements listed above compose nonmetallic inclusions in the form of mechanical admixture to magnetite. The NGR-spectrum of magnetite in the medium temperature associations corresponds, on the whole, to spectrum of stoichiometric magnetite. Small deflections of the NGR-spectrum parameters from stoichiometric composition observed in magnetite of the high temperature associations can be explained by the presence of the aluminium admixture. This agrees with the chemical analysis data [226].

The established features of composition and structure of the major constitutient ferromagnetic mineral of iron chert have resulted from variations of thermodynamic conditions in which iron chert has been formed. These have affected magnetic parameters of magnetite itself and of the entire rock.

Methods for investigations of the magnetic characteristics

As magnetic criteria characterizing stability of iron chert samples representing different facies against the high pressure and temperature effect the parameters I_n and I_{rs} are chosen. Of these, the natural remanent magnetization is more sensitive to a change in

PT-conditions. The I_n value and stability to the effect of demagnetizing factor are mainly dependent on the physical processes which have led to their formation. For metamorphic rocks the demagnetizing factors are, in the first place changes in pressure and temperature.

Magnetic parameters (natural remanent magnetization I_n and remanent magnetization of saturation I_{rs}) were studied under high hydrostatic pressure (to 2.5 kbar) on a special device designed at the Department of Physical Properties of the Earth Material, Institute of Geophysics, Academy of Sciences, Ukrainian SSR, Kiev [221].

Stability of I_n against the high pressure effect was studied by thermal demagnetization ($T = 20 \div 600$ °C) in the zero magnetic field.

I_{rs} was established for all the samples of the collection under study by using a constant magnetic field of 10000 Oe. In this case, the remanent magnetization reflects the magnetic state caused by almost the same strong magnetic field. The difference in the behaviour of $I_s = f(P, T)$ is, then, dependent mainly on composition of the constituent ferromagnetic mineral and its structure, etc. The latter is dependent on the conditions in which the mineral was formed and the degree to which these conditions have lasted up to the time of this study.

Experimental data

The results of the experimental studies of natural remanent magnetization I_n for iron chert under hydrostatic pressure to 2.5 kbar and high temperature are shown in fig. 1.

In iron chert sampled from the green schist facies wherein magnetite associates with the low-temperature minerals, the P effect on I_n is most clearly seen to 1 kbar. For a part of the samples (fig. 1 a, plot *1*) this P value has appeared sufficient to cause a complete loss of magnetization. Other samples (fig. 1 a, plots *3, 4*) loose more than 50% of the original I_n value within the same pressure range. A further increase of pressure lead to an insignificant decrease of the natural remanent magnetization of the rock.

Iron cherts sampled from the amphibolite facies with magnetite associating with the medium-temperature minerals are characterized by a wide range of I_n variations under pressure. So, at $P = 2.5$ kbar, the I_n/I_{n_0} ratio ranges from 0.19 to 0.83 (fig. 1 b, plots *5—11*).

The same limit values of the relative change in I_n/I_{n_0} under 2.5 kbar are characteristic of iron chert of the granulite facies wherein magnetite is associated with the high-temperature minerals (fig. 1 c, plots *5—7*).

Fig. 1 a, b, c show thermal demagnetization plots for the natural remanent magnetization of the iron chert samples under consideration. The samples show different stabilities of I_n against the high temperature effect. In samples with magnetite of the low-temperature mineral associations I_n has appeared the least stable (fig. 1 a, plot *2*), in those with the high-temperature magnetite — I_n has proved to be the most stable (fig. 1 c, plots *1, 2*). An intermediate position is occupied by the medium-temperature associations (fig. 1 b, plots *1—4*).

The low-temperature iron chert, which forms under *PT*-conditions corresponding to the thermodynamic parameters of the green schist facies ($T \approx 427$ °C, $P \approx 2$—5 kbar) [223, 224], is characterized by a fine- and micro-layered texture with the interlayer thickness measured by tenths of a millimeter. Such samples have small values of magnetic parameters and a low stability to the high temperature and pressure effects. They become

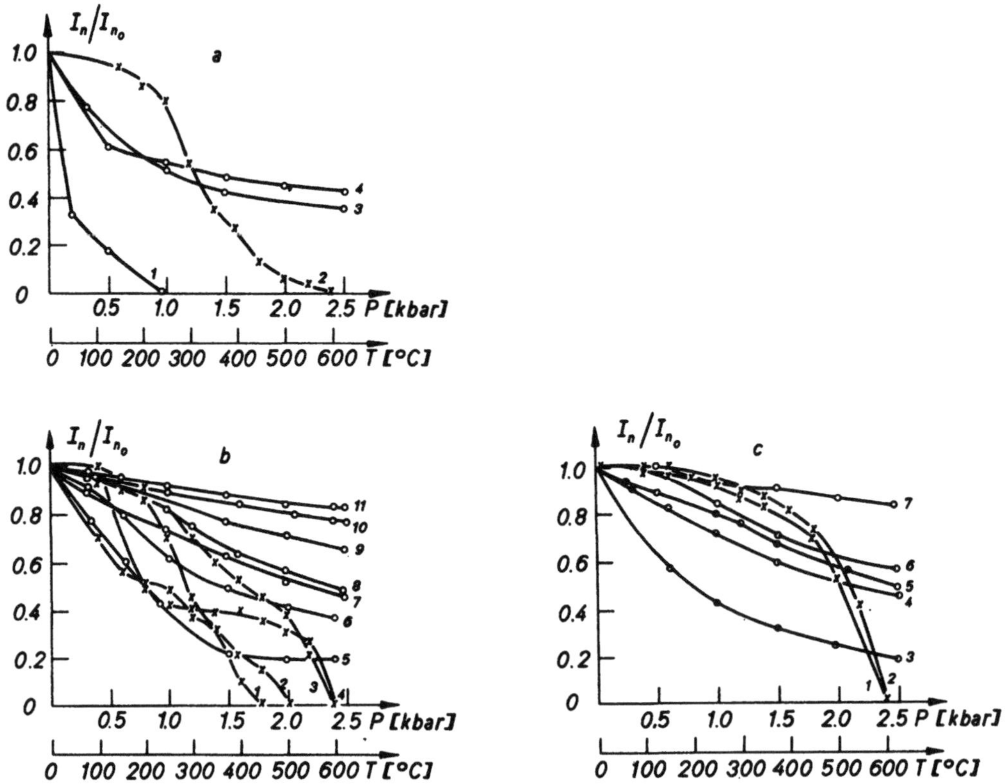

Fig. 1. The natural remanent magnetization (I_n) behaviour in quartzites under varying hydrostatic pressure (—○— *I*) and temperature (—×— *II*).

a — magnetite in association with low-temperature minerals; b — magnetite in association with medium-temperature minerals; c — magnetite in association with high-temperature minerals.

almost complitely demagnetized at $T \approx 300$—400 °C and $P \approx 1$—1.5 kbar (fig. 1a, plots *1*, *2*). This type of samples includes quartz carbonate magnetic quartzite containing magnetite of two generations. The first generation, as idiomorphic pure inclusions (without admixtures ones seen in microscope), composes ore intercalations; the second forms thin dispersed magnetite segregations in quartz interlayers. A low stability of I_n to pressure observed in samples containing magnetite of the first generations can be explained by the fact that, in the absence of nonmetallic inclusions, magnetite is a magnetically soft material.

Magnetization of some samples of iron chert from the green schist facies has appeared to be less responsive to the pressure effect (fig. 1a, plots *3*, *4*). According to mineralogical studies, these samples are composed of the carbonate hematite magnetite varieties. The presence of hematite in quartz intercalations, in groups with magnetite grains, is responsible for rigidity of the magnetic material to dynamic loading. With growing degree of metamorphism in ferruginous rocks under the conditions of granulite facies ($T \approx 627$ °C, $P \approx 3$—7 kbar) [224] their texture-structural features and mineral composition change and I_n values increase. The most stable to pressure is magnetization of the samples in which magnetite is associated with garnet, the mineral forming under high pressure. The aforementioned feature is clearly seen when the plots for $I_n = f(T)$ and $I_n = f(P)$ are compared (fig. 1b, plots *1*, *8*, *9*). With I_n sufficiently stable to pressure (I_n/I_{n_0} ratio is 0.82—0.87 at $P = 2.5$ kbar) some samples (fig. 1b, plots *2*, *4*) loose about 60% of the original I_n at about 250 °C and others (fig. 1a, plot *1*) are completely demagnetized at 450 °C. In certain samples (fig. 1b, plots *3*, *7*) a higher stability against both the demagnetizing factors is observed. The high stability of I_n against pressure found in a number of samples (fig. 1b, plots *5*, *6*) may be due to the fact that the constituent magnetite contains tiny oriented intergrowths of amphibole and biotite. This favours an increase of magnetic rigidity of the magnetic mineral. According to the mineralo-petrographic evidence magnetite, here was formed under the conditions of granulite facies.

The iron cherts, the conditions of whose formation correspond to the granulite facies of metamorphism ($T \approx 827$ °C, $P \approx 6$—10 kbar) [224], have mostly a high stability against the effect of pressure and temperature. The I_n/I_{n_0} ratios (at $P = 2.5$ kbar) range from 0.45 to 0.83 (fig. 1c, plots *4*—*7*). An exception is only plot *3* wherein $I_n/I_{n_0} = 0.18$. The shape of the $I_n = f(T)$ plots is indicative of a thermoremanent nature of I_n (fig. 1c,

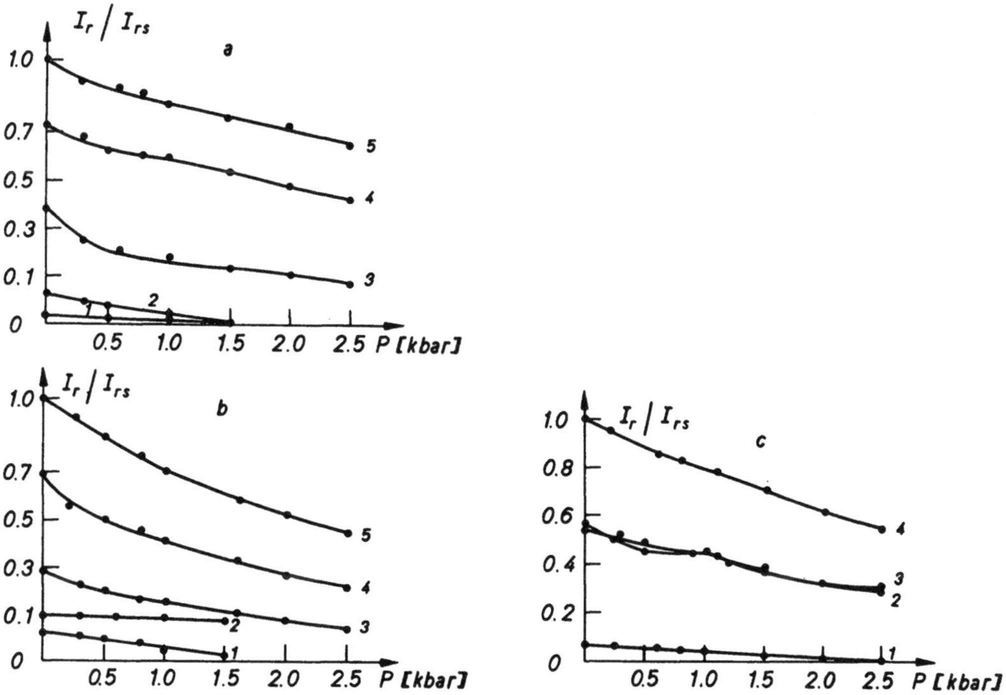

Fig. 2. Stability of normal remanent magnetizations (with $I_r \to I_{rs}$) of iron cherts under high pressure.
a, b, c — same as in fig. 1.

Legend: a) *1* — I_n; *2* — I_{r_1} ($H_e = 20$ Oe, $H_c = 1$ Oe); *3* — I_{r_2} ($H_e = 100$ Oe, $H_c = 4$ Oe); *4* — I_{r_3} ($H_e = 400$ Oe, $H_c = 14$ Oe); *5* — I_s ($H_e = 800$ Oe, $H_c = 16$ Oe).

b) *1* — I_{r_1} ($H_e = 40$ Oe, $H_c = 4$ Oe); *2* — I_n; *3* — I_{r_2} ($H_e = 100$ Oe, $H_c = 12$ Oe); *4* — I_{r_3} ($H_e = 200$ Oe, $H_c = 20$ Oe); *5* — I_{rs} ($H_e = 800$ Oe, $H_c = 31$ Oe).

c) *1* — I_{r_1} ($H_e = 40$ Oe, $H_c = 5$ Oe); *2* — I_{r_2} ($H_e = 100$ Oe, $H_c = 17$ Oe); *3* — I_n; *4* — I_{rs} ($H_e = 400$ Oe, $H_c = 32$ Oe).

plots *1*, *2*). The high-temperature association, to which magnetite in pyroxene varieties of iron cherts belongs, confirmes that magnetite has formed under the high-temperature conditions.

Consequently, the character of changes in natural remanent magnetization of iron cherts under high pressure and temperature corresponds to the *PT*-regime of their formation in the course of metamorphism. Higher values of P and T affecting rocks in the process of their formation lead to an increase of the I_n stability against the pressure and temperature effect during the experiment.

The influence of the conditions in which I_n forms on its stability against pressure can be observed in studying the process of magnetization of ferromagnetics. The shape of the normal magnetization plot may vary, even for one kind of a ferromagnetic (magnetite). It depends on the character of distribution of internal stresses, type of the domain boundaries, magnetic structure, the magnetoeleastic to other types of energy ratio, etc.

Fig. 2a, b, c shows the results of studies of the remanent magnetization stability (in different H_e fields) against pressure for the iron cherts containing magnetite associated with minerals of low-, medium- and high-temperature facies in particular hysteresis cycles. Here, indicated are the magnetic field intensity (H_e) at which normal remanent magnetization (I_r) is derived and the coercive force value (H_c).

For magnetite associated with the low-temperature minerals, the data presented show the possibility of deriving $I_n = I_r$ values from normal magnetization in the fields of the Rayleigh domain (fig. 2a, plot *1*). Stability of natural remanent magnetization I_n against pressure is commensurable with stability of I_r produced by a field of 20 Oe (plot *2*).

In the Rayleigh area the magnetization process runs mainly at the expense of a rotation of the 180°-domain boundaries which are not significantly affected by the field of inner stresses. In weak magnetic fields the rotation of these domain boundaries is primarily realized in certain spins. With the field intensity growing in the course of magnetization the domains forming not only 180° — but also 90° — and other boundaries become involved. The magnetization process is impeded by stresses of various kind including those requiring the external field energy to be overcome. The influence of dislocations on magnetization is due to heir magnetoelastic interaction with fields of domains and domain boundaries. The latter slows the movement of the domain boundaries and defines the value of coercive force. On approaching magnetic saturation the state of domain structure is dependent mainly on the effect of strong magnetic field causing a homogeneous magnetization of the ferromagnetic. Therefore the value and stability of magnetization to pressure are increasing at $I_r \rightarrow I_{rs}$. This is supported by data shown in fig. 2, plots *4*, *5*. The same general regularity in the $I_r = f(P)$ stability increase at $I_r \rightarrow I_{rs}$ holds for the results of the experiments with iron chert samples with magnetite associated with the medium-temperature (fig. 2b) and high-temperature (fig. 2c) minerals. The $I_n = f(P)$ plot related, to the former is placed between $I_{r_1} = f(P)$ and $I_{r_2} = f(P)$ plots. Normal remanent magnetization of the samples studied is obtained in fields of 40 and 100 Oe. In high-temperature minerals, stability of I_r obtained in a field of 100 Oe.

The aforementioned correlations between the value and stability to pressure of natural and normal remanent magnetizations can be observed only in experiments with strong magnetic fields. In the conditions of geomagnetic field the process of I_n formation in iron cherts occurs under the effect of high pressure and temperatures. The high *PT*-regimes which existed during formation of the iron cherts of medium- and high-temperature facies seem to be favourable for involving in magnetization the magnetorigid domains as well. Therefore natural remanent magnetization of considerable value and stable to the effect of external demagnetizing forces has been formed.

Remanent magnetization of saturation (I_{rs})

I_{rs} is usually used, along with other magnetic parameters, as a characteristics of composition and structure of the ferromagnetic component of the rock. Besides, I_{rs} of rocks undergoing regional metamorphism can provide information on high pressures and temperatures which have affected the rock.

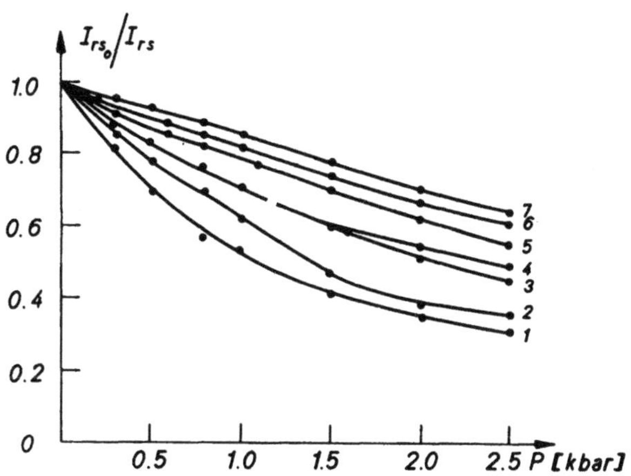

Fig. 3. The remanent saturation magnetization (I_{rs}) behaviour in iron cherts under high pressure.

Fig. 3 shows relative changes of I_{rs} in iron chert to 2.5 kbar. Monotony and uniformity of the I_{rs} decrease with pressure may be indicative of single-phase ferromagnetics. However, it is noted that $I_{rs} = f(P)$ plots are clearly varying with pressure in stability according to the adopted distribution of the rocks under study by facies.

In the course of the experiment, I_{rs} of the samples with magnetite associated with the high-temperature minerals appears to have the lowest stability in the low pressure range. For these samples a sharp decrease of magnetization with pressure to 1 kbar is characteristic (fig. 3, plots *1*, *2*). At higher pressures a gradient of decrease is lowing and the $I_{rs}/I_{rs_0} = f(P)$ curve is supposed to reach the asymptote, i. e. I_{rs} seems to stabilize. Such behaviour of I_{rs} is due to a prolonged effect of high thermodynamic parameters. The sudden fall of I_{rs} to 1 kbar may also be due to the preculiarity of formation of magnetite whose grains are almost devoid of visual inclusions. This circumstance, probably, favours the growth of gradients of the $I_{rs} = f(P)$ change.

I_{rs} of medium-temperature iron chert samples is more stable against pressure. In the initial P range small (about 30%) changes of I_{rs} are observed. Under higher pressure the gradient of I_{rs} decrease remains almost constant (fig. 3, plots *3*, *4*, *6*). This behaviour of I_{rs} under pressure can be explained by the effect of low pressures and temperatures on the rock formation in the conditions characteristic of amphibolite facies. The growth of I_{rs} stability against pressure is also due to the presence of oriented amphibolite and biotite inclusions in magnetite (fig. 3, plot *6*). I_{rs} of iron chert from green schist facies has appeared to be most stable under pressure, especially to 1 kbar. Here, the I_{rs} changes are as small as 15—20% (fig. 3, plots *5, 7*) and further P increase causes but an insignificant I_{rs} decrease, probably, due to the presence of various nonmetallic inclusions in magnetite grains.

As previously noted [49, 220], the I_{rs} stability against pressure is defined by internal stresses and changes in the domain structure of the ferromagnetic. Under compression, the process of rotation of the spontaneous magnetization vectors of certain domains is prevailing while the domain boundary movement is impeded by external stresses and displacement of dislocations. The influence of the above factors is especially strong to 1—1.2 kbar. With further growth of pressure a certain equilibration may take place resulting in a slower I_{rs} decrease and flattening out of $I_{rs} = f(P)$ plots.

Since the high pressure and temperature studies of I_{rs} deal with the change of almost the same magnetic state of the magnetite domain structure in different iron chert samples the data obtained can be used as an additional information on composition, structure and genetic features of magnetite. With pressure and temperature distribution within specific areas of deep crustal zones. available, the experimental and theoretical studies of remanent magnetization help to complete the characteristics of the change of magnetite-bearing rocks with depth.

Theoretical modelling

Modelling of changes in magnetic parameters of magnetite composing the deep-seated crustal rocks under the Ukrainian Shield has been performed previously [218].

The temperature change with depth calculated on the basis of the thermal field studies for the area of the Krivoy Rog basin [216] and the corresponding pressure distribution [16, 112, 225] were accepted as initial characteristics of the thermodynamic conditions of the region. The dependence of remanent magnetization of magnetite bearing rocks (including iron chert) on depth was then calculated from general analytical expressions [217—219], describing the change in magnetic parameters of mineral matter under high pressure and temperatures.

Fig. 4 shows relative changes in piezoremanent (a) and isothermal remanent (b) magnetizations of magnetite and magnetite-bearing rocks of Krivoy Rog against depth of their occurence. The characteristics of these changes under a simultaneous P and T effect have been obtained. The two remanent magnetizations are most responsive to changes in composition and structure of the ferromagnetic. Piezoremanent magnetization is of a special interest for it bears evidence of possible paleostresses and paleotemperatures undergone by rocks.

The calculations performed by a method worked out by the authors involve depths to 12—13 km (fig. 4a, b) since the possibility of interpolating the data on the changes of the magnetocrystalline anisotropy and magnetostriction constants is restricted to certain limits (4—4.2 kbar, 300 °C).

Fig. 4b sows the results of calculations of the $I_{pr} = f(H)$ dependence characterizing the behaviour of isothermal remanent magnetization I_{pr} of magnetite. A relative value of I_{pr} shows a rapid decrease to a depth of 3 km, then if increases almost in the same manner to 6—8 km. Below this depth its intensity starts falling more slowly. It is evident that the remanent magnetization values measured under normal conditions in samples cored from

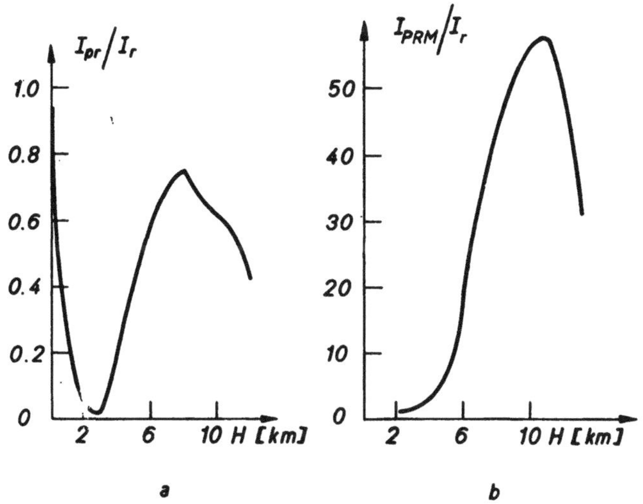

Fig. 4. The remanent magnetization change with depth (H) in magnetite composing the Krivoy Rog iron cherts (from the data of theoretical modelling)
a) isothermal remanent magnetization (I_{pr}); b) piezoremanent magnetization (I_{PRM}).

deep boreholes hould be systematically corrected for the effect of high pressures and temperatures.

Piezoremanent magnetization (I_{PRM}) of magnetite (fig. 4a) increases rapidly with depth to a maximum at a depth of 10—11 km below which it starts dropping rapidly. The dependence $I_{PRM} = f(H)$ is characterized by steep almost symmetrical rising and falling branches of the curve.

In maxima the values of paleostresses and paleotemperatures ($P = 2.9$ kbar, $T = 160$ °C) can be approximately estimated.

Conclusions

The results of the high pressure and temperature studies of magnetic properties of iron cherts show that the rate of the magnetic parameters change is essentially dependent on thermodynamic regime of the magnetite formation. Magnetites forming in different PT-conditions are distinguished by typomorphic features expressed in the size and internal structure of grains, the presence of mechanic inclusions and isomorphic admixture. As a result, magnetic properties of minerals and rocks under high pressure and temperature are usually steady within wide limits.

The iron cherts with magnetic associated with the low-temperature minerals is characterized, apart from a low stability of natural remanent magnetization to pressure and temperature, by a higher stability to pressure of remanent saturation magnetization. This is resulted from the difference in physical processes of the rock magnetization taking place in natural and experimental conditions. In natural conditions, the presence of inclusions and admixtures under relatively low PT-regimes of the rock formation involves in generation of natural remanent magnetization only a part of domains (mainly, low coercive). In laboratory, the presence of a strong magnetic field (of the order of 10000 Oe) removes the potential barriers produced by inclusions and favours the participation of all, or at least the majority of, the domains in the process of the rock magnetization. In demagnetization of the both types of remanent magnetization the changes of the domain structure reduce. However, to destroy natural remanent magnetization low and to destroy saturation magnetization — high pressures are required.

The iron chert with magnetite associated to medium- and high temperature minerals is characterized by an increase of the magnetite grains and decrease of the amount of mechanic inclusions and the presence of isomorphic admixtures. Stability of natural remanent magnetization depends, mainly, on the size, shape and relative position of magnetite grains within the rock conditioned by intensity of metamorphic transformations. Stability of remanent saturation magnetization is due to magnetic rigidity of magnetite dependent on the defects of its crystal lattice (at the expense of isomorphic admixtures).

On the whole, highly metamorphosed iron cherts, unlike the green schist facies formations, show a higher stability to *PT*-demagnetizing factors in experimental conditions.

The calculations of the magnetic parameters dependence on depth for magnetite bearing rocks (including iron cherts) have confirmed the necessity of indroducting corrections for the high pressure and temperature effect into the values measured in normal conditions. This is also rather important for interpretation of the anomalous geomagnetic field. Besides, from the behaviour of remanent magnetization with depth the values of paleostresses and paleotemperatures can be estimated.

The iron chert studies under conditions of varying thermodynamic regimes of the experiment have shown that application of pressure only is not sufficient for the iron chert demagnetization. This is possible only under a simultaneous effect of high pressures and temperatures.

THE HOPKINSON EFFECT OF MAGNETITE AND TITANOMAGNETITE UNDER UNIAXIAL COMPRESSION

M. JELEŃSKA

Introduction

The Hopkinson effect has recently been studied by many investigators [227—229], and many important features have been found. Some implications of the results of these investigations to the interpretation of magnetic anomalies and the state of magnetic minerals in rocks have been discussed.

The modelling of processes taking place in the rocks of the Earth's crust demands that we adopt thermodynamic conditions similar to those in the crust; this means that in addition to temperature we should introduce another parameter: pressure. Research into magnetic properties of rocks under high pressure has shown that the latter appreciable affects these properties.

The initial susceptibility is defined as the magnetization of a sample subject to so small changes of the magnetic field that the changes in magnetization are entirely reversible.

$$\varkappa_a = \lim_{\Delta H_i \to 0} \left(\frac{\Delta I_a}{\Delta H_i}\right)_{H_i=0,\, I_a=0},$$

where:

I_a — initial magnetization,

H_i — internal field

In general, the changes in magnetization take place either through the domain wall displacement or through the rotation of spontaneous magnetization vectors I_s. The initial susceptibility substantially depends on the domain configuration as well as the lattice defects and inclusions.

For the 180° Bloch wall displacement, the initial susceptibility of a cubic crystal with inclusions and internal stresses produced by the crystal deformation is given by [230]:

$$\varkappa_a \sim \frac{I_s^2}{C_1|\lambda|\sigma_{ex}K_1^{1/4} + C_2 K_1^{7/4} d^{3/2}},$$

where:

λ — magnetostriction constant,

σ_{ex} — external stress,

d — diameter of inclusions,

K_1 — crystalline anisotropy constant,

C_1, C_2 — constants which depend on the material.

Whence

$$\varkappa_a \sim \frac{1}{|\lambda|\,\sigma_{ex}}$$

$\varkappa_a$ will depend on temperature as $\dfrac{I_s^2}{|\lambda|\,|K_1|^{1/4}}$ in the presence of internal stresses and as $\dfrac{I_s^2}{|K_1|^{7/8}}$ in the presence of inclusions.

For the rotation processes, the initial susceptibility of a crystal with irregularly changing internal stresses σ_i has been estimated as follows

$$\varkappa_a \sim \frac{I_s^2}{\frac{9}{2}|\lambda|\,|\sigma_i| + 3|K_1|}$$

for $|\lambda|\,|\sigma_i| \gg |K_1|$

$$\varkappa_a \sim \frac{2I_s^2}{9|\lambda|\,|\sigma_i|} \sim \frac{1}{|\lambda|\,\sigma_{ex}} \qquad \text{if} \qquad \sigma_i \sim \sigma_{ex}\,.$$

Thus, in the region of domain wall displacements $\varkappa_a$ increases with increasing temperature as $\dfrac{I_s^2}{K_1^{1/n}}$, while in the region of rotations of the I_s vector it increases as $\dfrac{I_s^2}{K_1}$. Stresses, however, cause a decrease of $\varkappa_a$ as $\dfrac{1}{\lambda\sigma_{ex}}$ in the entire temperature range.

Experimental procedures

Igneous rocks, norites and basalts, containing titaniumrich titanomagnetite, magnetite and maghemite were investigated. The grain sizes ranged from few microns to millimeters, which corresponds to multi-domain grains.

The microscopic analysis of polished sections, X-ray investigations and measurements of Curie points yielded the following characteristics of the minerals under study:

1. Basalt containing nonoxidized titanomagnetite, with a Curie point of 100—160 °C, which corresponds to the 70—65% ulvospinel content. The grains are homogeneous and form two size classes: of 5—20 μm and 50—100 μm.
2. Basalt containing nonoxidized titanomagnetite with a Curie point of about 200 °C, which corresponds to the ulvospinel content of about 60%. Grain sizes of 3—40 μm.
3. Magnetite are containing magnetite with a Curie point of 575 °C. Magnetite forms clusters of crystals, in which single grains are hardly distinguishable.
4. Norite containing coarse-grained magnetite, partly oxidized to maghemite. Grains with diameters of about 2 mm, intersected by numerous ilmenite lamellae.

Because of the lack of appropriate instruments, we measured the magnetization in the 5—15 Oe magnetic field instead of the initial susceptibility. The value of the field depended upon the value of magnetization of the sample. The schema of the instrument has been described previously by [231]. We applied pressure of 600—800 kg/cm^2 parallel to the magnetic field direction.

Results

Fig. 1 shows the $I(T)$ curves obtained for the magnetic minerals described above.

As can be seen, the behaviour of $I(T)$ for the ore (sample 3, Fig. 1 a) is typical of MD magnetite. The Hopkinson peak height, 50%, also agrees with the values for such rocks given by other authors [227, 228]. The behaviour of $I(T)$ for norite (sample 4, Fig. 1 b) is more complex. The pronounced peak above 450 °C seems to be connected with the presence of magnetite, while the broad maximum around 300 °C may indicate the presence of maghemite.

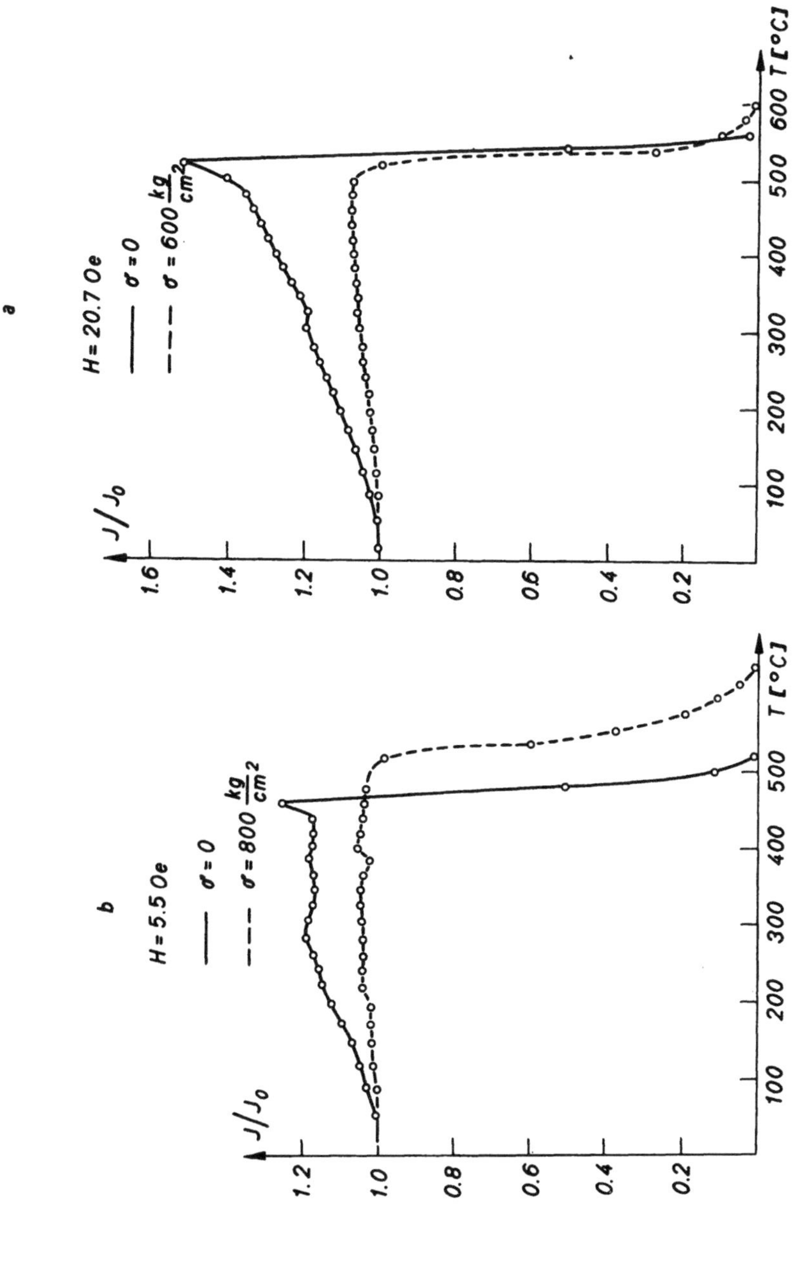

Fig. 1. Initial magnetization versus temperature: solid line — without compression; dotted line — under compression. a — Magnetite, sample 3; b — Magnetite oxidized to maghemite, sample 4;

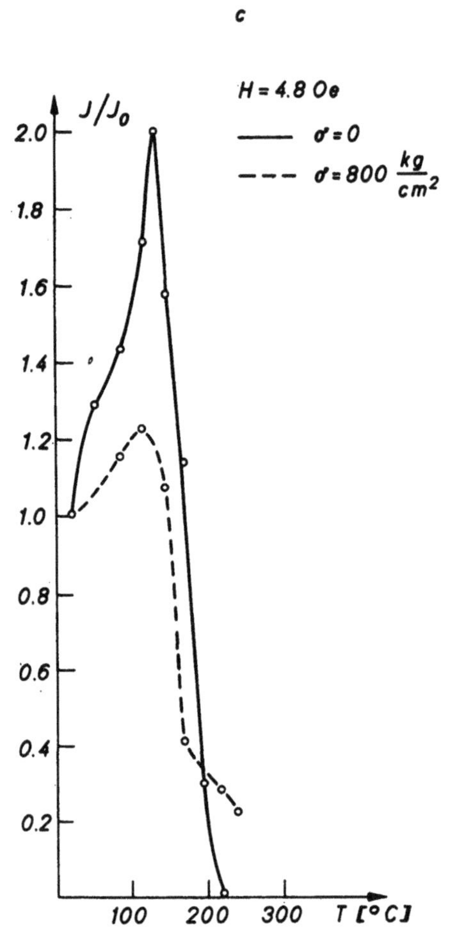

c — Titanomagnetite with Curie point of 200 °C, sample 2;

The $I(T)$ curve of the basalt containing titanomagnetite (sample 2, Fig. 1c) has a pronounced peak at 130 °C, i.e. just below the Curie point, which may indicate the presence of MD grains.

The basalt containing titanomagnetite (sample 1, Fig. 1d) with a Curie point of about 100 °C has the Hopkinson peak below room temperature. At the same time, its magnetization does not decrease sharply to zero. This points to the presence of another component of magnetization due to the small amount of partly exidized titanomagnetite. The titanomagnetite with a Curie point of 160 °C (Fig. 1e) has the Hopkinson peak at a temperature of abbut 70 °C. The other component of magnetization, which is due to partly oxidized material, is present also in this case.

The application of pressure changes the behaviour of curves $I(T)$ in two ways (Fig. 1). The increase of magnetization with temperature becomes smaller, in some cases being even absent (sample 3, norite — sample 4, basalt — sample 1). If present, the Hopkinson peak is less pronounced and displaced toward low temperatures. Another effect caused by pressure is the enhancement of oxidation processes. In basalt samples subject to pressure,

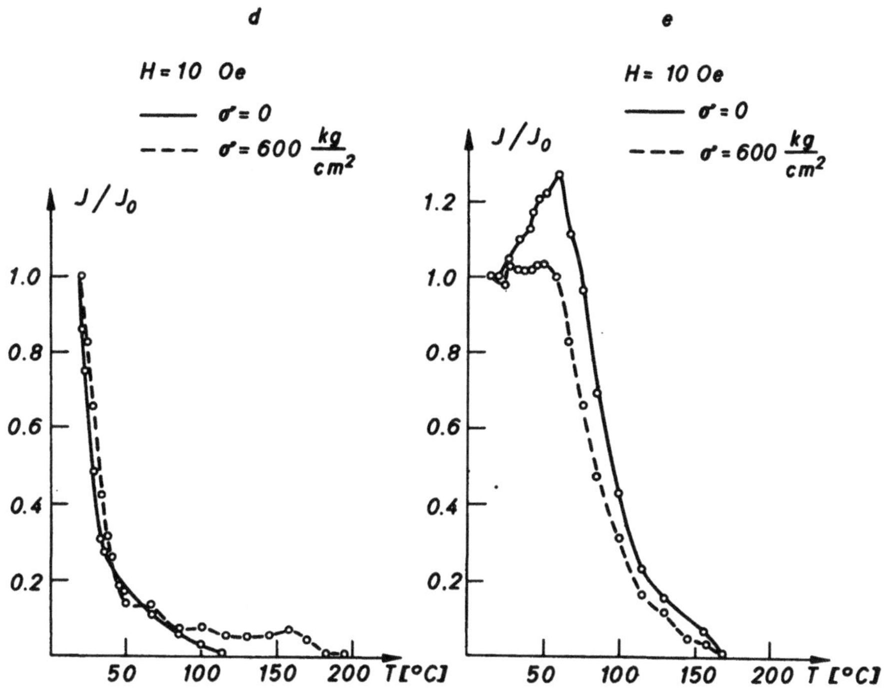

d — Titanomagnetite with Curie point of 100 °C, sample 1;
e — Titanomagnetite with Curie point of 160 °C, sample 1.

another component of magnetization appears in all the $I(T)$ curves. In magnetite subject to pressure we observe slight displacement of the Curie point. In norite, pressure causes the displacement of the Curie point of the main phase from 520 °C to 540 °C as well as the appearence of a small amount of another phase with higher Curie point (about 640 °C).

Temperature affects the Hopkinson effect in a similar manner (Fig. 2). The $I(T)$ curves measured for samples that had previously been heated to 600 °C at a rate of 1 °C/min also show a reduction of the Hopkinson effect and the appearance of another magnetic component. Temperature is more efficient in enhancing the oxidation processes than in reducing the Hopkinson effect. Pressure, on the other hand, is less efficient in its action on the oxidation of the material, producing much stronger decrease in the Hopkinson effect.

Similar effects are observed if a sample is compressed previously at room temperature. Basalt containing titanomagnetite (sample 2) was three times subject to pressure of 1000 kg/cm^2 at room temperature. Dependence $I(T)$ was measured afterwards. As can be seen in Fig. 2d, the Hopkinson peak is reduced and there appears another component of magnetization, indicating that oxidation took place.

Discussion

It follows from the obtained experimental curves $I(T)$ that:

1. The maximum on $I(T)$ curves becomes reduced and broadened under the influence of uniaxial pressure.

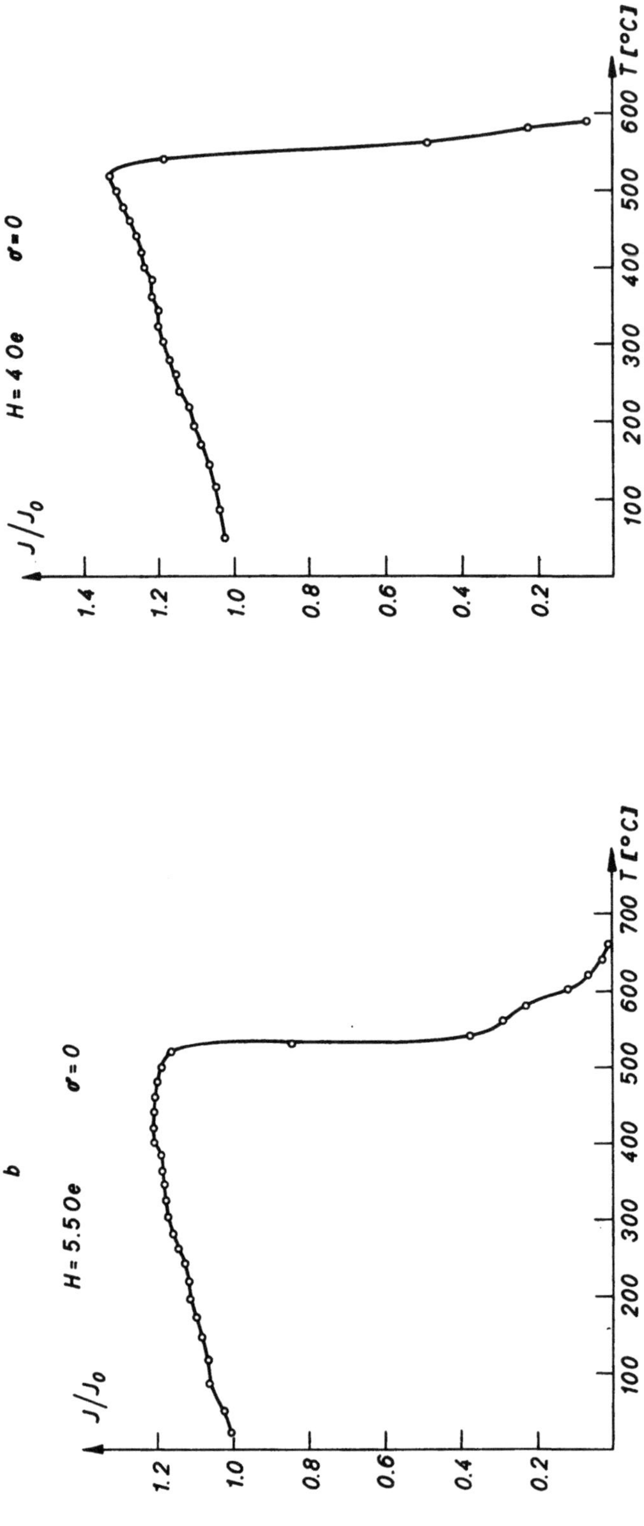

Fig. 2. Initial magnetization versus temperature:
a — Magnetite after heating to 600 °C, sample 3; b — Magnetite oxidized to maghemite after heating to 600 C, sample 4;

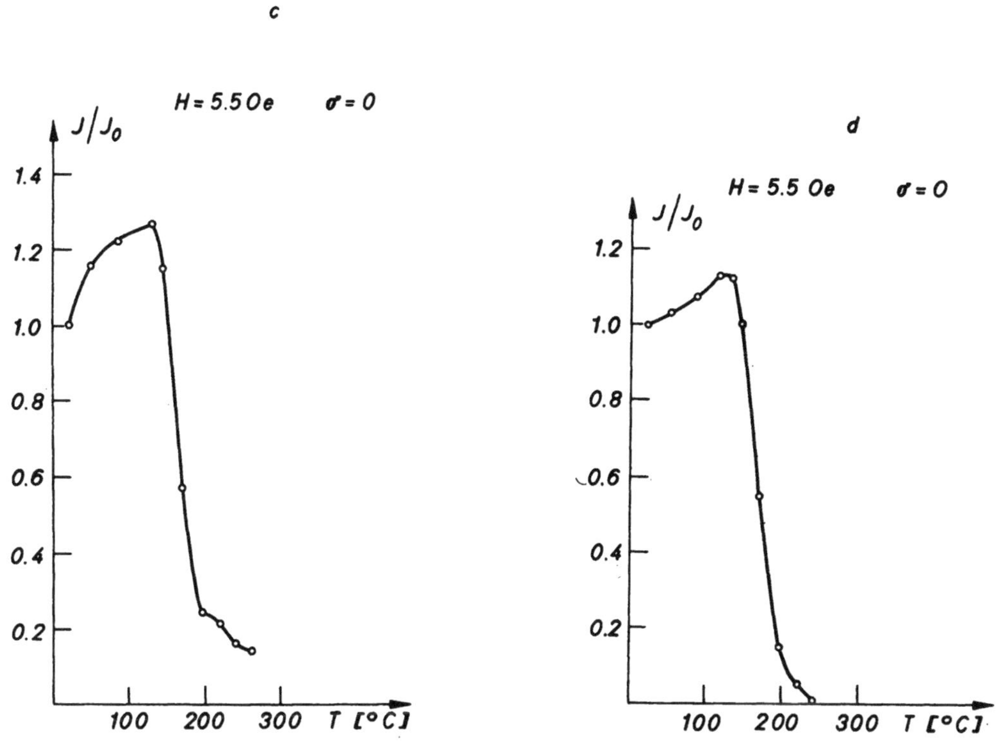

c — Titanomagnetite with Curie point of 200 °C after heating to 600 °C, sample 2;
d — Titanomagnetite with Curie point of 200 °C compressed to 1000 kg/cm² three times.

2. Pressure enhances the oxidation processes.
3. The behaviour of experimental $I(T)$ curves is generally consistent with theoretical dependences given previously [230].

According to the theory, the greater increase in susceptibility with increasing temperature should take place in the region corresponding to rotation processes, i.e. at high temperature, while in the region of domain wall displacement a little maximum should exist. Perhaps one should relate the intermediate maximum observed on $I(T)$ curves for norite (sample 4) and basalt (sample 2) to the domain wall displacement.

APPLICATION OF PULSE MEASURING TECHNIQUES WITH CYLINDRICAL SAMPLE SYMMETRY FOR DETERMINING THE THERMAL CONDUCTIVITY OF ROCKS UNDER HIGH PRESSURES

U. Seipold, W. Gutzeit, D. Stromeyer

Measurement of the thermal properties of rocks and minerals has been carried out much less intensively as compared to the elastic and electric properties. This is mainly due to the that there is no prospecting method directly based on the thermal data of field measurements. On the other hand, there is a great interest in more detailed information on the thermal parameters on the part of geophysics. For the development of theoretical conceptions and models describing the time behaviour of the thermal processes in the Earth and the planets, the determination of the corresponding material parameters is, of course, a prerequisite. For determining the heat flow, the combination of field measurements with laboratory investigations is also indispensable. All these applications require a knowledge of the thermal properties under the conditions of greater earth depths, viz., under extreme pressure and temperature conditions. Naturally our knowledge in this field is even less satisfactory than for normal conditions. In carrying out the experimental investigations, some fundamental problems are encountered which are caused by the special conditions of generating high pressures and temperatures:

1. In the pressure chamber only a small volume is available for the sample and the measuring equipment to be mounted in its immediate vicinity (heater, temperature sensor). For the quasi-hydrostatic generation of pressure the volume to be expected amounts to only a few millilitres; in case of hydrostatic chambers, it is in the 100 millilitres order of magnitude.
2. Since the pressure-transmitting medium must always be in close thermal contact with the sample, it is difficult to realize a thermal insulation.
3. The connection between the sample and the components of the measuring equipment within the pressure chamber on the one hand and the external measuring equipment on the other can be accomplished only by means of electric cables. Leading-in these cables through the chamber walls presents considerable problems, therefore their number should be as small as possible.

In view of the above facts it appears useful to test non-steady methods for measurements under high pressure. This paper shall especially investigate the possibilities of pulse techniques. In the past few years pulse techniques have found increasing application for the determination of thermal quantities, especially also for measurements at very high temperatures [232, 233, 234]. Therefore it suggests itself to try to apply these methods also at high pressures. However, the frequently used generation of thermal pulses by means of laser radiation [235, 236] should be realizable only in special cases, because it would require the installation of high-transmissivity components (windows or transparent plugs) in the pressure chamber. Therefore the pulse generation will largely be accomplished by means of electric heaters. An essential aspect lies in the selection of the sample geometry to be used. Simple sample shapes are encountered in case of linear and radial heat flows. In these cases the mathematical treatment of the problem is also relatively uncomplicated. Particularly for cylindrical samples the radial heat flows is in rather good agreement with the geometrical conditions in most of the high-pressure chambers. For linear heat flow it is difficult to limit thc lateral heat losses (vertical to the direction of propagation). The close contact of the sample with the pressure-transmitting medium makes an efficient insulation impossible. In most cases the available space does not suffice for installing a protective heating system.

When a cylindrical sample is used, there are two variants:

1. A heating wire (line source) arranged in the axis of the sample is used; the temperature curve observed after heating up by an electric pulse is recorded as a function of time by means of a thermocouple at a certain distance a.
2. The tubular heater encloses the cylindrical sample. The temperature sensor is arranged in the axis. Thus in this configuration the thermocouple and the heater are exchanged as compared with the former variant.

The results of theoretical treatment are in effect similar in both cases. For the evaluation of measurements it is necessary to know the temperature distribution as a function of locus and time after a pulse had acted upon the system at the time $t = 0$. The pulse is supposed to be of arbitrarily short duration; the medium, i.e. the sample, is supposed to be homogeneous, isotropic and of a arbitrary extension. Certainly these assumptions are only conditionally satisfied. Therefore, later on, we shall investigate what sorts of effects are introduced by restrictive conditions.

The calculation of the temperature distribution is based on Fourier's equation of heat conduction, where the certainly correct further assumption is made that there are no sources or sinks of heat in the medium. Then the following relation holds:

$$\nabla^2 T - \frac{1}{K}\frac{\partial T}{\partial t} = 0\,,$$

where K denotes the thermal diffusivity. At the time $t = 0$ a heat pulse, the time behaviour of which is described by a Dirac delta function, is introduced at the point $\vec{r} = \vec{r}'$. Then according to [237, 238], for $t \geqq 0$ and all points $\vec{r} \neq \vec{r}'$ the solution of the equation of heat conduction has the form

$$T(x, y, z, t) = \frac{q}{\varrho \cdot c(4Kt)^{3/2}} \cdot \exp\{-[(x - x')^2 + (y - y')^2 + (z - z')^2]^2/4Kt\}\,.$$

In what follows, we shall consider a cylindrical heater, the axis of which coincides with the z-axis of the coordinate system, its centre coinciding with the origin. Let the heater have a radius a and a length $2l$ extending from $z = -l$ to $z = +l$. The surface area $F = 4\pi al$ of the heater can be decomposed into elements of area $\mathrm{d}F = a\,\mathrm{d}\varphi\,\mathrm{d}z'$. Each element of area can be considered a point source producing the infinitesimal portion $q = \mathrm{d}Q = Q\,\mathrm{d}F/F$ of the total thermal energy Q introduced. The resulting temperature distribution is obtained by summation over the effects of all point sources, i.e. by integration over the total heater:

$$T = \frac{Q}{\varrho \cdot c} \cdot \frac{1}{8(\pi K t)^{3/2}} \cdot \frac{1}{4\pi l} \int_0^{2\pi} \int_{-l}^{+l} \exp\left[-(a^2 + z'^2)/4Kt\right] \mathrm{d}\varphi \, \mathrm{d}z' .$$

$$T = \frac{Q}{\varrho c} \cdot \frac{\exp(-a^2/4Kt)}{8\pi K t l} \cdot \operatorname{erf}\left[l/(4Kt)^{1/2}\right].$$

From the condition $\left.\frac{\mathrm{d}T}{\mathrm{d}t}\right|_{t_m} = 0$ for the extreme value,one obtains a conditional equation for the thermal diffusivity:

$$\frac{a^2}{4Kt_m} = 1 + \frac{l \cdot \exp(-l^2/4Kt_m)}{2(\pi K t_m)^{1/2} \cdot \operatorname{erf}\left[l/(4Kt_m)^{1/2}\right]} = 1 + \varepsilon(l, K).$$

The correction term $\varepsilon(l, K)$ takes the values 0.5 and 0 as $l \to 0$ and $l \to \infty$, respectively. It is easily shown that for the range $0 < l < \infty$ one has a monotone decreasing function. The above conditional equation can be solved by means of successive approximation. The first approximation is obtained by means of the setup $\varepsilon_1 = 0$:

$$K_1 = a^2/4t_m .$$

If the experimental conditions are selected so as to satisfy $l/a \geqq 2.15$, then the relative error is already less than 1 %. In case of a less satisfactory l/a ratio, a second approximation can be obtained by using the value K_1 to calculate a correction term $\varepsilon_2 = \varepsilon(l, k_1)$ and inserting the latter into the above conditional equation for the purpose of calculating K_2. This second approximation should be sufficient for all practically occurring cases. The first approximation is the exact solution in cases where an arbitrarily long heating element is used. By inserting the expression for K_1 into the relation for the temperature T_m at the time t_m, a relation results from which it is possible to calculate the specific heat c:

$$\varrho \cdot c = \frac{Q}{2l \cdot e \cdot \pi \cdot a^2 \cdot T_m} .$$

If one assumes that the relation between the thermal diffusivity and the thermal conductivity

$$\lambda = \varrho \cdot c \cdot K$$

is valid for rocks, then it is also possible to calculate the thermal conductivity:

$$\lambda = \frac{Q}{8l \cdot e \cdot \pi \cdot t_m \cdot T_m} .$$

The relations for the reciprocal experimental arrangement (line source with a thermocouple arranged at a distance a) are derived analogously. An arbitrary number of point sources is thought to be arranged along the z-axis in the interval from $-l$ to $+l$ and their effects in the test point are summed up. For the temperature distribution one obtains in this way

$$T(r, t) = \frac{Q}{8\pi l \cdot K \cdot t \cdot \varrho \cdot c} \cdot \exp\left(-r^2/4Kt\right) \cdot \operatorname{erf}\left[l/(4Kt)^{1/2}\right]$$

taking into account the cylindrical symmetry.

If test points are considered, the source distances of which are small as compared with the length of the heater, i.e. for which $l/r \geqq 2$, then especially for the measuring point $z = 0$, $r = a$ the above relation reduces to

$$T(t) = \frac{Q}{8\pi l \cdot K \cdot t \cdot \varrho \cdot c} \exp\left(-a^2/4Kt\right).$$

Again the extremal condition $\left.\frac{\mathrm{d}T}{\mathrm{d}t}\right|_{t_m} = 0$ yields a relation for the thermal diffusivity:

$$K = a^2/4t_m .$$

Inserting this value into the relation for the temperature distribution, it is also possible in this case to determine the specific heat,

$$c = \frac{Q}{2\pi e \cdot l \cdot a^2 \cdot \varrho \cdot T_m}$$

as well as the thermal conductivity,

$$\lambda = \frac{Q}{8\pi \cdot e \cdot l \cdot t_m \cdot T_m} .$$

An essential advantage of the described pulse methods lies in the fact that in principle it is possible to determine the temperature and thermal conductivities by one and the same measurement.

In the derivation described above, the assumptions included idealized conditions which cannot completely be realized in practice. Therefore it has to be investigated how these deviations from the assumptions will influence the results. In what follows, we shall first consider the influence of the finite pulse length [239].

Let the function describing the time behaviour of temperature at the measuring point for zero pulse length be denoted by $T'(t)$. As shown above, the thermal diffusivity is derived from the extreme condition

$$\frac{\mathrm{d}T'}{\mathrm{d}t} = 0 \quad \text{has the form} \quad K = \frac{a^2}{4t_m} .$$

The temperature as a function of time for a finite pulse duration Δt reads as follows:

$$T(t) = \int_0^{\Delta t} T'(t - \tau)\,\mathrm{d}\tau .$$

The extreme condition can be rearranged in the following way:

$$\frac{dT(t)}{dt} = \frac{d}{dt} \int_0^{\Delta t} T'(t-\tau)\, d\tau = \int_0^{\Delta t} \frac{d}{dt} T'(t-\tau)\, d\tau = 0 .$$

In view of $\frac{d}{dt} T'(t-\tau) = -\frac{d}{d\tau} T'(t-\tau)$, it follows that

$$\frac{dT(t)}{dt} = \int_{\Delta t}^{0} \frac{d}{dt} T'(t-\tau)\, dt = \int_{t-\Delta t}^{t} \frac{d}{d(t-\tau)} T'(t-\tau)\, d(t-\tau),$$

$$\left.\frac{dT}{dt}\right|_{t_0} = T'(t_0) - T'(t_0 - \Delta t) = 0 .$$

Hence the points $(t_0 - \Delta t)$ and t_0 must lie on different sides of the maximum of $T'(t)$, i.e. on different sides of t_m. The distance between t_0 and t_m is less than Δt. One has

$$(t_0 - \Delta t) < t_m < t_0 .$$

Consequently, the desired value $K = a^2/4t_m$ satisfies the following inequality:

$$\frac{a^2}{4t_0} < K < \frac{a^2}{4(t_0 - \Delta t)} .$$

For the sample dimensions used by us, the measuring times t_m and t_0 are in the order of magnitude of a few seconds for most of the rocks. According to the above derivation there is no significant error introduced into the result by the finite pulse duration, if the latter is small compared with these values, i.e. if it is in an order of magnitude of a few hundredths of a second. To satisfy this condition does not involve any significant restriction.

In case of investigations under high pressure it is expected that the sample volume decreases. Since the sample dimensions directly enter into the conditional equations for K, λ and c, for substances having a relatively high compressibility it is necessary that the variation of distances with pressure is taken into account.

A second assumption of the theory presents much greater problems. In the derivation, an infinitely extending medium was assumed. Because of the limited volume in the pressure chamber this condition is far from even being an approximate realization. The actual situation lies between two limiting cases:

1. The sample is bounded by a medium having a very high conductivity ($\lambda \to \infty$).
   In this case the transport of heat in the sample is forced as compared to the situation with arbitrary extension. The maximum is reached more rapidly, so that the measured conductivity is too high. Moreover the maximum is lower.
2. The sample is bounded by a medium having a very low conductivity ($\lambda \to 0$).
   Then the thermal energy is accumulated in the sample. The temperature of the maximum is high, the extremum is broad and is reached with a time lag. The measured conductivity is too low.

When the measurements in the pressure chamber are carried out, the pressure-transmitting medium surrounds the sample. For a quasi-hydrostatic generation of pressure, pyrophyllite is used in most cases. Its thermal properties do not differ much from those of most of the rock samples.

For a closer investigation of the influence exerted by the surrounding medium, one and the same cylindrical sample was embedded in a number of thermally highly different media. Fig. 1 shows the temperature curves recorded at the measured point. It turns out that the above qualitative expectations are confirmed. As an empirical rule, the evaluation showed that the measured value of thermal diffusivity is proportional to the logarithm of the thermal conductivity of the surrounding medium. In carrying out high-pressure investigations, the absolute value of the thermal conductivity of the samples is of less interest than the relative variation of this quantity as the pressure is increased. According to the hitherto gained experience this variation is not very considerable. In our opinion the measuring error caused by the influence of the surrounding medium can be accepted under these conditions.

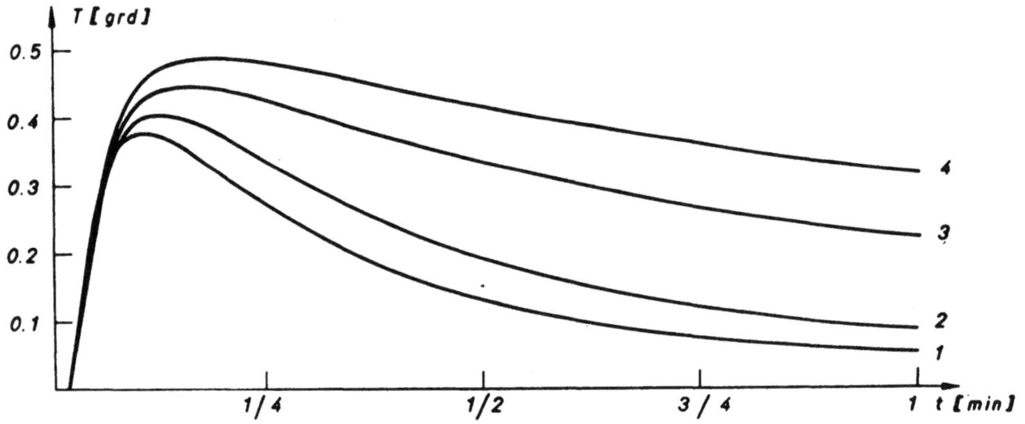

Fig. 1. Temperature curves for different media surrounding the sample.

The pulse measuring technique was practically tested in a quasi-hydrostatically operating pressure generation plant, viz., a solid pressure-transmitting medium was used. Thus it is possible to reach higher pressures than with hydrostatic detectors. Moreover, measurements performed at a high pressure and simultaneously at a high temperature can be realized in a less complicated way. It is this feature which we consider as being a special advantage of the presented pulse measuring technique, especially in the variant with a line-shaped heater. By means of an external tubular heater it is possible to regulate the temperature of the sample in the pressure chamber. For measuring the heat conductivity, a thermal pulse can independently be introduced by means of a line-shaped heater in the sample axis. In principle only one thermocouple is required for the measurement.

The forces necessary for pressure generation are produced by means of a triaxial compressive-strength tester made by VEB Werkstoffprüfmaschinen, Leipzig. This equipment generates maximum forces of 300 Mgf (600 Mgf in the vertical direction) in three mutually perpendicular directions. By means of a special die system these forces are transmitted to a 23 × 23 × 34-mm cuboid. This cuboid consists of pyrophyllite and contains the sample to be investigated in a cylindrical bore. The first experiments were carried out using a tubular heater made of graphite. The cuboid faces are greater than those of the compression dies, the dimensions of which are 20 × 30 mm and 20 × 20 mm, respectively.

Thus in the initial phase of each compression test pyrophyllite flows into the gaps between the dies, thus sealing of the pressure chamber. The further deformation of these gaskets enables the sample to be further compressed as the applied force increases, thus increasing the pressure within the sample. Consequently, the pyrophyllite acts at the same time as a pressure-transmitting medium and as a sealing material. The dimensions stated above represent a compromise. In case of a very large oversize of the cuboid as compared with the die faces, a great quantity of material flows into the gaps between the six tapered compression dies. Consequently, a great fraction of the available force is required for compressing the gaskets, so that only a relatively small part acts on the sample itself. On the other hand, if the oversize of the cuboid is too small, then only very thin gaskets develop; in this case as well a great fraction on the force is required in order to achieve a certain deformation, although the areal extension of the gaskets is smaller.

The electrical connection of the heater and the thermocouples is accomplished by means of contacts leading to the dies. In this variant the risk of lead interruption by shearing-off during the compression is relatively low. The possibility of threading the wires through the sealings has proven much more critical. Fig. 2 shows the sample arrangement.

The thermal pulse is generated by a current surge fed through the graphite heater. The required high-current pulse is supplied by discharging a bank of capacitors. It has a

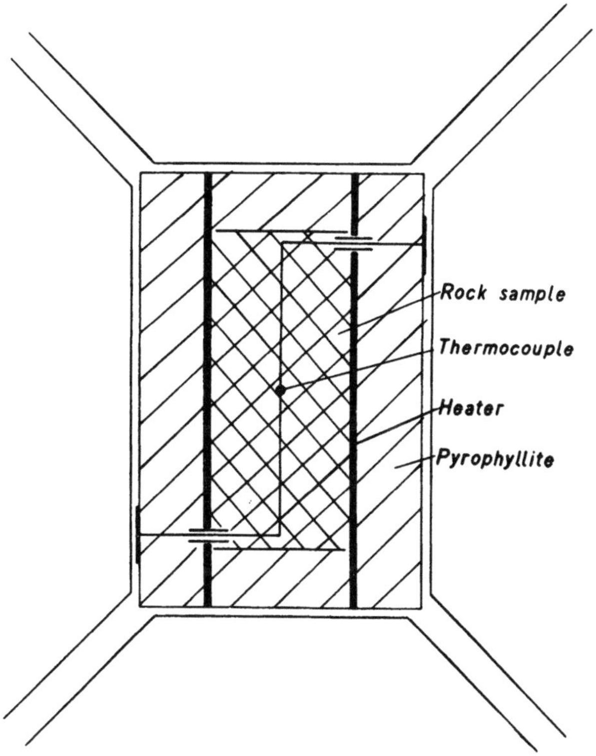

Fig. 2. Sample arrangement.

duration of several hundred microseconds. Long transient phenomena are restricted by a suitable design of the supply leads. The time behaviour of temperature in the centre of the sample is detected by means of an Ni-NiCr thermocouple and recorded by means of a sensitive potentiometer tape recorder. The pulse delay time t_m can be read off from the record with sufficient accuracy. Here it proves advantageous that a voltage surge is induced in the thermocouple by the magnetic field of the electric current surge for generating the thermal pulse. This voltage surge is clearly observable on the record and allows an accurate determination of the starting point of the pulse delay time.

In certain cases the maximum of temperature is relatively flat. Then the end point of the pulse delay time is not defined with sufficient accuracy. Under these conditions it has proven useful to use the time $t_{1/2}$ required for reaching half the maximum temperature T_m for calculating the thermal diffusivity. Because of the steeper slope of the record in this range, the time $t_{1/2}$ corresponding to the temperature $T = \frac{1}{2} T_m$ can be determined with much higher accuracy than the time t_m corresponding to the maximum at $T = T_m$. The equation required for calculating the thermal diffusivity is the same except for a constant factor:

$$K = 0.373 \frac{a^2}{4t_{1/2}}.$$

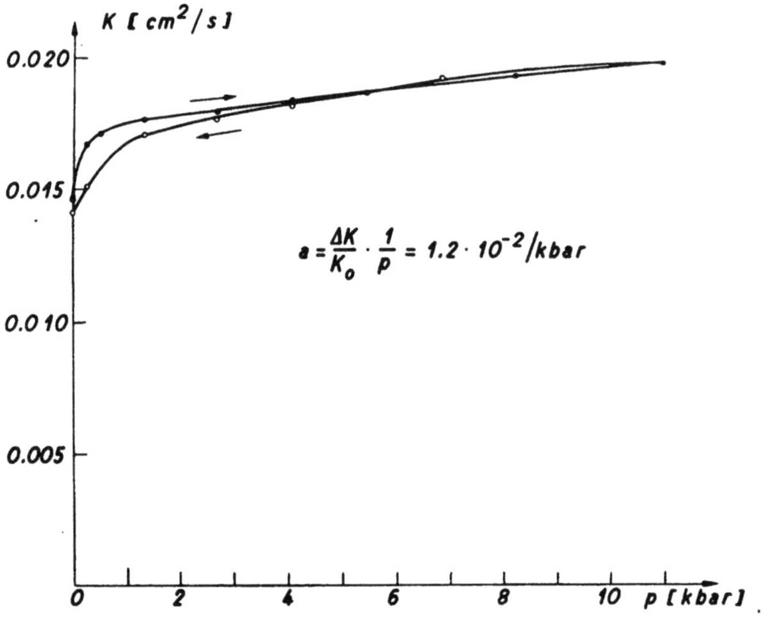

Fig. 3. Measurements of temperature conductivity in a sandstone sample.

By way of example, Fig. 3 shows results of measurements of the thermal diffusivity in sandstone with increasing and decreasing pressure. The absolute value and the variations caused by a variation of pressure are of the same order of magnitude as in previous measurements performed by the authors with similar rocks in a hydraulic pressure chamber by means of a modified Angstroem technique [240]. With the new setup it was however possible to cover a much greater pressure range.

CHAPTER V

Physical Properties of Lunar Rocks

THEORETICAL AND EXPERIMENTAL INVESTIGATIONS OF PHYSICAL PROPERTIES OF LUNAR ROCKS UNDER EXTREME THERMODYNAMIC CONDITIONS

R. WÄSCH and R. ENGLER

Theoretical Investigations

Elastic properties

Young's modulus of rocks

For rock samples of the Apollo missions 11, 12, 14, and 15, the velocities of the longitudinal and shear waves were measured as functions of pressure [169, 170, 171].

Comparable investigations of terrestrial samples were carried out by many authors [6, 8, 9, 13, 35, 57, 64, 78, 79, 80, 92, 93].

The comparison of wave-velocity measurements of lunar and terrestrial rock samples under pressure showed that the wave velocities in lunar rocks exhibit a much higher pressure dependence than that shown in terrestrial rocks, the density and the chemism being similar in both rock types.

First of all, the moduli of elasticity E were calculated for the samples 12002 (Apollo 12), 14310 (Apollo 14) and 15418 (Apollo 15) as functions of pressure (up to 6.5 kbar) according to equation (1) below [171 a]:

$$E = \frac{v_s^2 \cdot \varrho(1 - 2\sigma)(1 + \sigma)}{1 - \sigma}, \tag{1}$$

where

$$\sigma = \frac{1}{m} \quad \text{and} \quad m = \frac{2(v_l^2/v_s^2 - 1)}{v_l^2/v_s^2 - 2}; \tag{2}$$

v_s — velocity of shear waves,
v_l — velocity of longitudinal waves,
σ — Poisson's ratio,
m — lateral contraction number.

Several measurements were carried out with the sample 12002. For calculating the modulus of elasticity, mean values of v_l and v_s were obtained. These results are shown by Fig. 1 as a function $p = f(E)$.

Two functions approximately describing the curves were derived from this plot:

$$p = f(\ln E), \tag{A}$$
$$E = E_{max}(1 - k_0\, e^{-\varkappa P}); \quad K < 1,$$
$$E_{max} \text{ — maximum value for } p \to \infty,$$
$$E = f(\ln p), \tag{B}$$
$$P = A_0\, e^{\varkappa E} - A_1 .$$

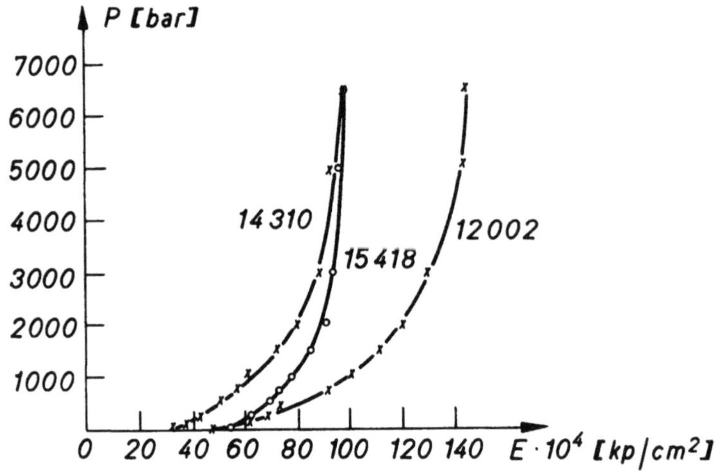

Fig. 1. Young's modulus E as a function of pressure p for lunar rocks.

With the boundary conditions $P = 0$, $E = Ea$ one obtains the expression

$$P = A_0(e^{\varkappa E} - e^{\varkappa Ea}) .$$

It turns out that the function stated under (B) provides a better description of the situation.

Poisson's ratio

Table 1 lists the Poisson's ratios (eq. 2) for the samples 12002, 12022, 14310, and 15418 for increasing pressure. It is observed that σ does not change with pressure, vit., σ represents a constant. Consequently, the observed behaviour differs from that described by [172] for porous rocks.

Table 1. Poissons's ratio of lunar rocks with increasing pressure.

P [bar]	sample 12002	12022	14310	15418
0; 1	0.28	0.36	0.31	0.25
100	0.282	0.362	0.303	0.25
250	0.307	0.35	0.31	0.26
500	0.309	0.36	0.3	0.286
750	0.309	0.35	0.31	0.283
1000	0.3	0.29	0.33	0.283
1500	0.296	0.32	0.299	0.282
2000	0.298	0.34	0.299	0.285
3000	0.298	0.33	0.295	0.282
4000	—	—	0.296	0.287
5000	0.29	0.31	0.296	0.282
6500	0.307	0.3	—	—

From this it can be concluded that the lunar rocks were subjected to a different compaction process. This supports Schoen's presumption that the pressure-dependent variations of wave velocity are caused by disturbances in the inner bonds of the rocks.

The quantity σ can be interpreted as a constant for the Apollo 12 and Apollo 14 samples. For sample 15418, the Poisson's ratio increases with increasing pressure in the lower pressure ranges up to 1 kbar. This is analogous reported by [172].

Compressibility of the lunar samples

The compressibility as well as the porosity provides further information for a comprehensive description of the lunar matter. Thus it is possible to make qualitative statements about the strength of the essential minerals of the moon.

First of all, the compressibility $\varkappa$ was determined as a function of pressure for the samples 12002, 14310, and 15418 according to equation (3).

$$\varkappa = \frac{3(m-2)}{Em}. \tag{3}$$

The results are summarized in Fig. 2.

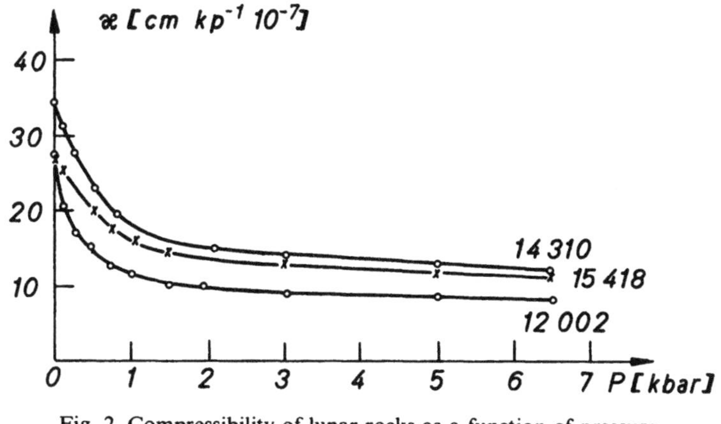

Fig. 2. Compressibility of lunar rocks as a function of pressure.

In Fig. 2 it is observed that for the samples 12002 and 14310 up to a pressure of 2 kbar, the $\varkappa$ values decrease to between one half and one third of their initial values and approach a constant value as the pressure is further increased.

Sample 15418 shows a similar behaviour; however, for pressures below 2 kbar the decrease is not as rapid as above. These statements are in good agreement with the measured results of [173] and [174].

Investigated is in [173] the compressibility of lunar crystalline rock, micro-breccias, and the dust fraction of Apollo 11 under pressures upt to 40 kbar and at room temperature. All samples are porous. Fig. 3 shows the compressibility of the three above-mentioned rocks.

It was found that the compressibility of the lunar matter is typical for porous terrestrial rocks. Thus in this case an analogy between lunar and terrestrial matter can be

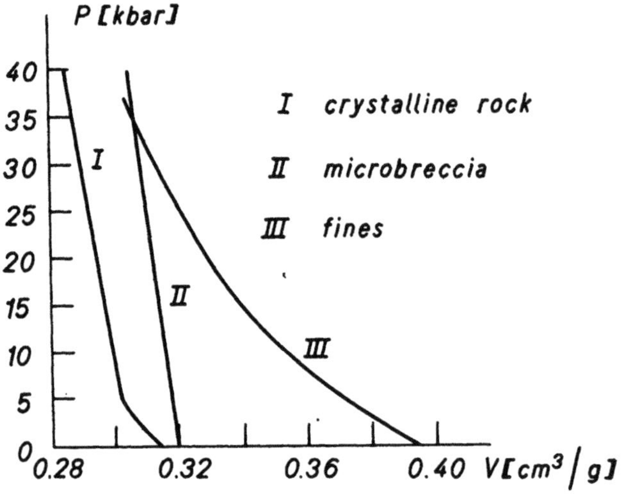

Fig. 3. Compressibility of Apollo 11 rock samples as a function of pressure (according to Stephens [173]).

observed. By way of example, the volume variation of crystalline rock amounts to 9.8% (Fig. 3, curve *I*) at 40 kbar. This result is similar to that found on many terrestrial eruptive rocks. The lunar samples have an initial porosity of more than 5%, the samples being slightly porous even under the highest pressures.

At 35 kbar the compressibility is 1.5 Mb^{-1}. For this pressure range the value of compressibility is very high. The cause of the higher compressibility of the lunar rock lies in its residual porosity.

This leads to the conclusion that all samples including the crystalline rock contain pores up to a pressure of 40 kbar [173]. In [174] are performed compressibility measurements of the rock sample 10017 with pressures up to 6 kbar. The sample 10017 (Apollo 11) represents a finegrained rock. The measured volume compressibility values of the lunar samples are equivalent to the compressibility values of terrestrial rocks; the compressibility rapidly decreases from a very high value as the pressure is increased up to 1 kbar (see Fig. 4).

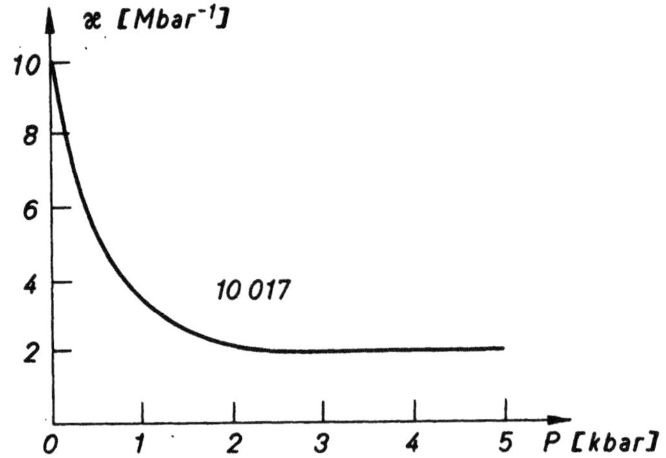

Fig. 4. Compressibility of the Apollo 11 rock sample 10017.

This effect is due to fissures enclosed in the rock. It was found that above 0.5 kbar the linear compressibility is almost istotropic which, among other things, is an indication of a low degree of positional orientation of the minerals in the rock. It indicates however a certain positional orientation of the fissures. This preferred orientation of the fissures, could for example, have been caused by impact or thermal shocks.

Thermal investigations

Thermal expansion

In comparison with terrestrial rocks, the low value of the longitudinal- and shear-wave velocities and the thermal conductivities at pressures below 100 bar as well as the high compressibility values for several Apollo 12 basalts show that the lunar rocks are extremely fractured.

Hence the coefficient of thermal expansion of lunar rocks is slightly below that of the terrestrial rocks; at a depth of more than 90 km, it reached a value of

$$21 - 23.5 \times 10^{-6} \text{ deg}^{-1}$$

due to the closing of fissures.

On the other hand, the value of the thermal expansion coefficient of the terrestrial rock is as low as

$$16 \times 10^{-6} \text{ deg}^{-1} .$$

Specific heat

The specific heat c_v is directly related to the thermal expansion α; for solids the following equation (4) holds:

$$c_v = \frac{\alpha}{\varrho \varkappa \gamma}, \tag{4}$$

where ϱ denotes the density, $\varkappa$ the compressibility and γ the Grueneisen ratio. For different rocks of the Apollo 12, Apollo 14 and Apollo 15 missions, the values of specific heat were calculated according to equation (4). The c_v values are listed in Table 2. The corresponding value for terrestrial olivine

$$c_v = 10 \text{ kgf cm g}^{-1} \text{ deg}^{-1}$$

Table 2. Specific heats of various lunar rocks.

sample number	12002	12022	14310	15418	
main element	Albite	Olivine	Anorthite	Anorthite	
α	17	25	15	15	10^{-6} deg^{-1}
γ	0.5	1.2	0.7	0.7	
$\varkappa$	2.8	28	34.5	27.5	10^{-7} cm^{-2} kgf^{-1}
ϱ	3.32	3.3	2.88	2.8	g/cm^3
c_v	3.64	2.25	2.22	2.79	kgf cm g^{-1} deg^{-1}

shows that the specific heat of terrestrial minerals is much higher. The c_v data stated in Table 2 refer to normal pressure. Calculations of the specific heat under termal influence have already been carried out, whereas c_v was not yet determined as a function of pressure. In what follows, $c_v = f(p)$ shall be determined according to equation (4). For that purpose two assumptions can be made:

1st assumption α constant
$\varkappa$ variable
2nd assumption α variable
$\varkappa$ variable
ϱ constant for one sample

The results derived from the first assumption are shown in Table 3. Using. the second assumption, the behaviour of the specific heat under pressure for a variation of compressibility and of the thermal expansion shall be investigated. Up till now no measured values of the thermal expansion as a function of pressure were stated in the literature.

Table 3. Specific heat as a function of pressure (1-st assumption).

P [bar]	12002	14310	15418
0	3.64	2.22	2.79
100	4.75	2.37	3.06
250	6.04	2.68	3.25
500	6.45	3.23	3.72
750	8.3	3.75	4.22
1000	8.55	4.49	4.92
1500	9.83	4.49	4.92
2000	10	4.95	5.44
3000	10.7	5.29	5.75
4000	—	5.44	5.8
5000	11.9	5.86	6.28
6500	12.3	—	—

c_v [kgf cm g^{-1} deg^{-1}]

The determination of $\alpha = f(p)$ was carried out using the model [175], Fig. 5, and the pressure vs. depth survey according to [171]. Table 4 summarized the results derived from the second assumption. Table 3 lists the results derived from the assumption that the thermal expansion is constant and the compressibility variable. For the rock samples of the Apollo 12, 14 and 15 missions one obtains a continuous increase in the specific heat with increasing pressure. Only at a pressure of 2 kbar does c_v approach a constant value. The second assumption makes the trasition to a constant value appear even more distinctly and thus permits an explanation of the discrepancy between the specific heats of the terrestrial and lunar rocks. From Table 4 it can be seen that the c_v values of the lunar rock initially increase, whereafter a stabilization occurs.

Thus it must be supposed that in a pressure range between 1 and 1.2 kbar the fissures close and then also the extremely fractioned lunar surface rock behaves like a material having no significant elastic disturbances and is comparable with terrestrial rocks.

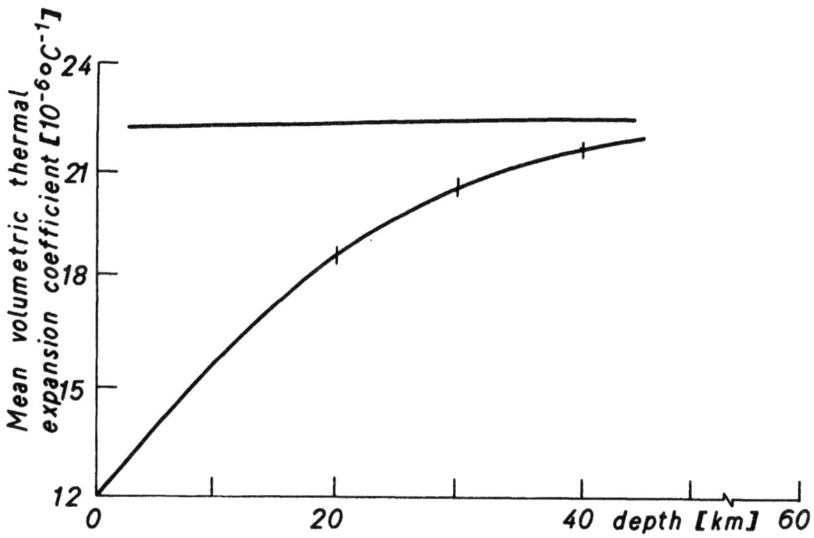

Fig. 5. Mean coefficient of thermal volume expansion as a function of depth for the lunar rock sample 10057 (Apollo 11) according to Baldridge und Simmons [175].

Table 4. Specific heat as function of pressure (2nd assumption).

P [bar]	$\varkappa$ [10^{-7} cm^2 kgf^{-1}]	c_v [kgf cm/g deg]	α [10^{-6} deg^{-1}]
0	28	2.58	12
380	16.5	5.7	15.5
570	15	6.8	17
760	12.4	9.0	18.5
1000	12.0	9.78	19.3
1270	11.2	11.0	20.4
1460	10.5	12.0	20.9
1650	10.38	12.5	21.5
1840	10.31	12.7	21.7
2030	10.2	12.8	22.0

α from Baldridge and Simmons [175]

Experimental Investigations

X-ray investigations at high pressures

The hitherto published results of mineralogic-crystallographic and physico-chemical investigations of lunar minerals support the view which was generally taken already in the past that the properties of the rock-forming minerals occuring on the moon in effect coincide with those of terrestrial analogues.

The experiments provide however no information on whether this agreement continues to exist when, for example, the individual physical parameters are investigated under the influence of pressure. In the present case the structural behaviour of lunar minerals under different pressures as compared to their terrestrial analogues was of interest. The objects of these investigations were olivine and anorthite fragments, and the experiments were carried out using a diamond double-piston-apparatus (Fig. 6).

The principle of construction of this chamber is based on a conception of pressure generation which was described for the first time by [176]. The pressure is generated by means of a spindle via a helical spring and two levers. The sample is arranged between two diamond pistons with a compression area having a diameter of 0.35 mm and irradiated in the direction of compression. Diffraction patterns obtained in this way are identical with Laue transmission patterns; they are recorded on cut film. This method does not place any high requirements on sample preparation. By means of a suitable adhesive the crystal fragments are slightly stuck to the working face of a diamond piston in order to prevent them from

Fig. 6. Diamond double-piston high-pressure chamber ready for service for the investigation of crystalline structures at pressures up to 100 kbar.

sliding off the working face when the pistons are pressed together. Thus a "powder film" having a maximum thickness of 0.1 mm develops.

By way of example, Fig. 7a and 7b show diffraction patterns of lunar and terrestrial anorthites. Here, on each photograph, the line spectrum at normal pressure (nP) is contrasted with the spectrum at about 100 kbar (hP). Naturally, the quality of the diffraction patterns is not comparable with that of photographs obtained by the conventional technology; the individual lines can however readily be identified. It is clearly observed that the patterns obtained at normal pressure and at 100 kbar are almost identical. A slight diffuse broadening occurs in the high-pressure pattern for a few reflexes.

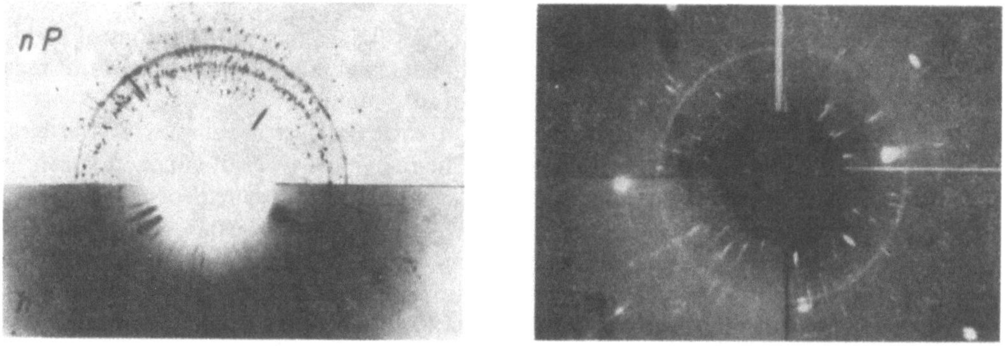

Fig. 7. X-ray diffraction patterns of a lunar (Luna 20) (a) and a terrestrial (Italy, Vesuvius (b) anorthite at normal pressure (nP) and 100 kbar (hP).

Consequently, on the basis of this experiment it can be stated that under the influence of a uniaxial pressure ranging from normal pressure to 100 kbar (other parameters were not varied), the crystal structures of lunar olivine and anorthite crystals do not change; only slight lattice distortions are indicated. Corresponding terrestrial minerals show an identical behaviour. The diffraction patterns show the same picture. The high-pressure X-ray investigations in this case confirm the results of other cristallographic experiments. The identity of lunar and terrestrial crystal structures also extends over their properties under high-pressure conditions. The different conditions of origin and the variety of influence to which the minerals were subjected since their formation on the Moon and on the Earth, do not cause any significant modifications of their behaviour in this case.

Investigation of the electrical resistivity of regolith as a function of pressure and temperature

From a great number of papers published in the literature on measurements of the electrical resistivity of rocks under different thermodynamic conditions, it is known that the results of such investigations depends on many factors. Above all, these factors include the mineralogic composition, the structure (grain size, grain contacts, voids), moisture and anisotropy. Moreover, the methods of investigation influence this result. Therefore it is not possible to perform an exact classification of the rocks on the basis of resistivity measurements [177, 178]. Effusive rocks, metamorphic rocks, and part of the sediments are mainly composed of highly insulating minerals; therefore, the electrical resistivity of these rocks is much higher than that, for example, of ore-bearing rocks. On the other hand, it is relatively independent of the mineral composition, because most of the rock-forming minerals have about the same electrical resistivity.

In spite of these considerable restrictions with respect to the classification of rock types on the basis of electrical measurements, it is however possible to perform rough subdivisions. Thus acid rocks generally have a higher electrical resistivity than basic ones. This can be explained by the fact that the essential mineral constituents of acid rocks (quartz, feldspar, etc.) have a higher electrical resitity than the main constituents of basic rocks (olivine, pyroxenes, etc.).

However, at this point we shall once more point out that even small additions of highly conductive substances (ores, certain oxides, sulphides, and the like) and moisture may reduce the electrical resistivity by several orders of magnitude.

The mechanisms of conduction occurring in a rock are just a complex as the rock itself. The rock in effect appears as a dielectric. The individual types of conduction can be determined only by experiment, and it is possible to use the temperature dependence as well as the Hall effect as criteria [177].

As already mentioned, a great number of factors are involved in the investigation of real rocks. Add to this that with increasing temperature and pressure, reactions between individual rock components must be taken into account to an increasing extent. Furthermore, it shall not be ignored that, for example, the low-temperature contributions of ionic and electronic conduction are subject to large variations and for each sample depend on its thermal history. Moreover, the activation energy of the movement of atoms at grain boundaries is certainly lower than that of atomic movements at lattice defects in rock crystals, but its respective contribution to the total conductivity cannot be determined satisfactorily.

As it was already pointed out, the electrical conductivity of dielectrics greatly depends on the temperature. This temperature dependence is also most suitable for determining the nature of conductivity in a material. The relationship between the conductivity (σ_e) and the temperature (T) is relatively clear, if there is only one type of conduction or one value of the activation energy (E). Of course for rocks this is not the case. In most cases the type of conduction changes with increasing temperature, because then charge carriers having a high activation energy become inportant. If the material to be investigated is highly affected with "impurities", the charge carriers associated to the latter are less strongly bound than, for example, lattice ions, and an increased conductivity occurs already at lower temperatures. Then the produced interstices can also more readily be occupied by lattice ions. For these reasons, both the low-temperature and the high-temperature contributions must always be taken into account in the investigation of conductivity.

For the experimental investigation of the electric conductivity of regolith, several milligrams of the Luna 16 and Luna 20 matter were used in each case. The samples were not separated and were taken from the respective dust fractions. Preparation and measurement were carried out under normal atmospheric conditions, viz., at ambient air humidity.

For carrying out the mesurements by means of a special device, the lunar dust was pressed into thin pellets (diameter — 4 mm, thickness — 0.15 mm), which during the measurement were arranged in a ring of synthetic mica (Novomicanite). The sample prepared in this way was placed between the pistons of a Bridgman double-piston apparatus with external

Fig. 8. Bridgman double-piston apparatus with external heating system for measuring the electrical resistivity of rocks at high pressures and temperatures.

heating system (Fig. 8). The measurement itself was performed according to the 2-electrodes method, where the high-pressure pistons at the sam time acted as electrodes. Pressure and temperature were varied between 1 or 10 and 25 kbar and between room temperature and 300 °C; under these conditions the electrical resistivity was measured. The results are shown in Fig. 9. The units of measurement in the coordinate system are suitably chosen, so that the activation energy (E) can readily be determined from the curves.

For the Luna 16 regolith, specific electrical resistivity ranging from 3.4×10^{-6} (ohm · cm)$^{-1}$ at 1 kbar and room temperature and 8.3×10^{-6} (ohm · cm)$^{-1}$ at 26 kbar and 300 °C were determined. The corresponding values for the Luna 20 regolith are between 3.7×10^{-6} (ohm · cm)$^{-1}$ at 1 kbar and room temperature and 8.6×10^{-6} (ohm · cm)$^{-1}$ at 26 kbar and 300 °C. The activation energies calculated from the measured values amount to several hundredths of an electron volt for both samples.

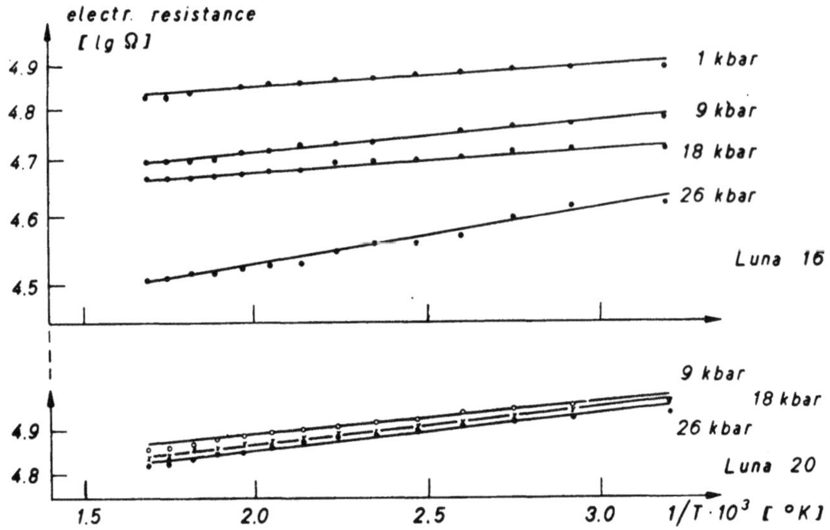

Fig. 9. Electrical resistivity of the Luna 16 and Luna 20 regolithes (dust fraction) as a function of pressure and temperature.

If the results are compared with data of other authors, who performed similar investigations on comparable rocks, then the here determined values (those of the electric conductivity as well as those of the activation energy) are in part much less than those stated in the literature.

In [179] is investigated the electric conductivity of the sample 14063 (Apollo 14) under normal atmospheric conditions at termperatures ranging from 20 to 300 °C. In this temperature interval the conductivity of the breccia varied from 10^{-12} (ohm · cm)$^{-1}$ to 10^{-6} (ohm · cm)$^{-1}$. From [180] are known the conductivities of the Apollo 11 and Apollo 12 samples 10024,22, 10048,53, and 12053,47. The material consists of compact basaltic micro-breccias. The measurements carried out in a helium atmosphere provided values between 10^{-10} and 5×10^{-5} (ohm · cm)$^{-1}$ in the temperature interval between 20 °C and 300 °C. The same author [180] further investigated the Apollo 15 samples 15555 (porous basalt) and 15418 (anorthositic breccia). At room temperature and in a helium atmosphere, values of 10^{-12} — 10^{-11} (ohm · cm)$^{-1}$ for sample 15418 and 10^{-11} — 10^{-9} (ohm · cm)$^{-1}$ for sample 15555 were determined.

From all these investigations an activation energy of 0.3 to 0.8 eV can be derived.

The electric conductivity of terrestrial rocks has already been investigated for large collections and a great variety of rock types. For comparison purposes Table 1 lists some values used by [178] for similar investigations. The conductivities were investigated under normal pressure and temperature conditions.

The data listed in the table illustrate the large scatter of even the measured values. However, it also demonstrates the high degree with which the electric conductivity depends on factors such as the humidity. In most cases for rocks having a certain water content it is increased by serveral orders of magnitude as compared with dry rocks.

Furthermore, a statistical analysis of large series of measurements confirms the fact that, on the average, basic rocks have a higher conductivity than acid ones. In these cases as well, the individual ranges show very large overlappings.

Table 5. Electric conductivity of rocks under normal conditions [177]

		Conductivity (ohm · cm)$^{-1}$	
		humid (natural)	dry
Acid rocks:			
Granite	Azerbaidshan	3.3×10^{-8}	—
Granite	Kola peninsula	6.2×10^{-9}	3.3×10^{-16}
Feldspathic porphyr	Australia	2.5×10^{-6}	—
Leukophyre (Albite)	Georgian SSR	3.4×10^{-5}	2.5×10^{-9}
Basic and ultrabasic rocks:			
Diabase	Georgian SSR	2.6×10^{-5}	3.0×10^{-8}
Basalt	Armenian SSR	6.25×10^{-6}	—
Periodites		3.3×10^{-6}	1.5×10^{-6}

For the above-mentioned reasons an interpretation of the electric conductivities measured in this case (in Luna 16 and Luna 20 regolithes) is very difficult. Add to this, moreover, that the applied experimental method has a great influence on the respective result of measurement (this is also true for all other investigations). Therefore direct comparisons with results of other authors are always extremely problematic.

As it was already mentioned, the samples consisted of lunar rock dust. One way however assume that already at a pressure of a few kilobars the individual particles form a sample compacted to such a degree that it is quite comparable with compact rocks. If the pressure is further increased (to about 5 kbar), then probably most of the pores and fissures contained in the individual fragments are largely closed. This is likely to be the main cause of the relatively high conductivity of the investigated samples, in contrast to results of other authors (cf. [180], [179]. Since the samples were already pre-compressed by high pressures before the experiments were started, the influence of grain boundaries (contact resistivity at the grain boundaries, grain contacts, fissures etc.) is likely to be largely suppressed. This assumption is supported by the fact that, for example, in the case of the Luna 16 matter the conductivity values at 1 and 26 kbar do not significantly differ from each other. Since the experiments were carried out under normal ambient conditions, the samples also had a certain moisture content. However, this moisture can be contained only in small traces, because the conductivities also show only a relatively slight temperature dependence.

This conjecture is supported by the fact that in each case the samples were very small (a few mg) and the individual measuring cycles dragged on over many hours. Accordingly at 300 °C most of the moisture should be evaporated out of the sample. Other investigations show, to which high degree the conductivity of the rocks is influenced by their water content (see for example [177]). Thus is was found that a water content as low as 0.1% already causes the electric conductivity of rocks to increase by several orders of magnitude. The general influence of temperature is known and has been verified by many experiments. In most cases, however, it clearly shows up only at temperatures of several hundred degrees centigrade. The temperature of maximum 300 °C applied in the experiments carried out here certainly are not high enough to activate a greater number of charge carriers, and the slight increase in conductivity is likely to be explained mainly by compaction effects in the sample as a consequence of the applied pressure.

The values of the activation energy, which amount to serveral hundredths of an electron volt in contrast to several tenths of an electron volt as found by other authors for similar substances, are considerably lower than initially expected. In this case, the sample composition probably plays an important part. It is known that the regolith, and here particularly the Luna 16 matter, contains other constituents in addition to the dominant minerals, such as oxides and in some cases also small traces of metals. A similar statement is true in the case of the Luna 20 matter. Since, due to the comstitution and the small quantity of the available matter, the samples were not separated, it is possible that the relatively high conductivity and the low activation energies are at least partially due to such admixtures. For the Luna 20 sample this fact is likely to have a greater importance than for the Luna 16 matter, because the Luna 20 anorthosite as an acid rock should, as a matter of principle, have a lower electric conductivity than the Luna 16 "basalt".

CHAPTER VI

Geophysical Consequences and Applications

INFLUENCE OF PRESSURE AND JOINTS ON THE VELOCITY OF ELASTIC WAVES IN JOINTY ROCKS IN THE CASE OF EARTHQUAKES

H. STILLER, F. C. WAGNER, and H. VOLLSTÄDT

Wave velocities in seismic regions and fracturing processes

For a long time past, the petrophysical investigations of the behaviour of elastic waves in rocks under pressure and in the fracturing region, respectively, have indicated variations of wave velocities showing up as both increases and decreases, and hence pointed to the possibility that in the regions of seismic centres similar varitions of wave velocity occur. Now relevant investigations [241, 242, 243] in fact proven that pressure bursts and earthquakes are preceded by considerable variations of wave velocity, and with come certainty it can be assumed that the variations of wave velocity are premonitory phenomena indicative of earthquakes. The results of the evaluations of low-intensity earthquakes [244] and of the San Fernando earthquake of Febr. 9th, 1971 show that in the course of days and months (in the case of low-intensity earthquakes or swarm earthquakes (for high-intensity earthquakes) the wave velocities decrease by 10—20%, and shortly before the outburst of an earthquake return to their original values. For the San Fernando earthquake, the decrease in wave velocity began in 1967 and reached its maximum during several months in 1968. A minimum in the curve of the ratio v_p/v_s appears just as distinctly.

The explanation of this phenomenon requires stressdependent and time-variant changes of the matter condition during the fracturing process. Fracturing is no sudden, short-time event, but the fracturing process must be considered to include the fraturing phenomena in the material structure which preceded the actual parting rupture. On the basis of fracturing investigations in limestone, in [246] was concluded that fracturing takes place in several phases:

a. The dilatancy phase
   Micro-fractures start developing at stresses below the breaking value.
b. Inclusion phase
   Considerable quantities of micro-fractures develop at stresses close to the breaking value.
c. Closure phase
   A micro-fracture starts growing, and as a consequence of the changed stress conditions the remaining micro-fractures close.
d. Growth phase
   The previously closed fractures again open, new fractures develop and the micro-fracture increases up to a critical length.

By [247, 248] is supposed that probably a delayed inflow of liquid into the developing fractures occurs. In this way the elastic bonds are inproved and the wave velocity, which decreases during the fracturing process, will again increase. The finally increasing pore pressure is believed to initiate the rupture. Consequently, apart from this hypothesis of liquid redistribution, which appears some what risky, the causes of these variations of wave velocity obviously can be found in the variations of fracture structures. The dimensions and quantities of fractures must taken an optimum distribution in order to effect a maximum variation of wave velocity. [249] likewise consider varying fracture distributions to be the cause of the wave velocity variations.

The elastic moduli are decreased [250—252] also in the case of closed fractures, i.e., of fractures having no volume of joint, but an internal friction of the fracture surfaces.

Measurements of wave velocity in porous model bodies

The decrease in wave velocity during the formation of numerous micro-cracks can be compared to ultrasonic wave-velocity measurements in porous model bodies with given pore sizes [253].

For the experimental determination of the influence of the pores on the wave velocity in porous rocks, porous test specimens with a well-defined and uniform pore size are required. Those materials which are in principle the most suitable ones for geophysical investigations, i.e. porous sediments, are useless due to their undefined pore size. Therefore rock-like specimens were made of gypsum with embedded substances of high air content (foamed-plastic particles or activated coal). The substances were separated into fractions of different sizes by means of sieves, and one fraction each was added to the cubical gypsum specimens having an edge length of 25 mm. The investigated samples had the following pore diameters (in mm): <0.06, 0.06—0.09, 0.09—0.12, 0.12—0.25, 0.25—0.30, 0.30—0.50, 0.50—0.60, 0.60—0.75, >0.75. The porosity always amounts to 1.3%, i.e., the number of pores decreases with increasing pore diameter. The ultrasonic frequency is 4 MHz.

In Fig. 1 the points show the measured values of wave velocity, v_p. The non-porous gypsum exhibits the highest wave velocity; after adding the pore substance, the wave velocity decreases of approximately 10% as referred to the wave velocity in particle-free gypsum.

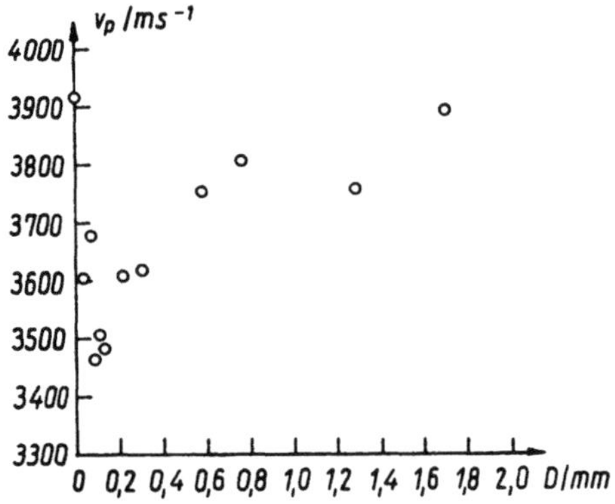

Fig. 1. Wave velocity v_p in samples with pore diameter D.

Formulae for the description of the wave-velocity variation

For the analytical description of the variations of wave velocity mentioned at the beginning and taking into account the existing pressure effects, it is possible to use two equations which were derived from wave velocity vs. pressure investigations of terrestrial as well as lunar rock samples and which are based on the consideration of fractured or porous materials. Investigated is in [254] the dependence on pressure of the propagation velocity of elastic waves in lunar and terrestrial rock samples and arrives at the result that the much higher pressure dependence of lunar rock samples cannot be attributed either to variations of the elastic properties of the or to variations of the pore and joint volumes, respectively. Rather it must be assumed that these great variations of wave velocity are the result of incomplete elastic bonds in a rock having no appreciable volume of joint. As a measure of such disturbances, a quantity K_l can be introduced which is defined as follows: in an arbitrary cross-section of length l, let a total length $K_l \cdot l$ be characterized by incomplete bonds of the constituents. The assumption that the decrease of K_l with increasing pressure is proportional to the quantity K_l itself, i.e., that

$$\mathrm{d}K_l = K_l A_l \,\mathrm{d}P \qquad \text{or} \qquad K_l = K_{0l}\,\mathrm{e}^{-A_l P}\,, \tag{1) (2}$$

leads to a relation for the velocity,

$$v_p/v_F = (1 - K_{0l}\,\mathrm{e}^{-A_l P})^{1/2}\,, \tag{3}$$

where K_{0l} is the initial value of the quantity K_l, A_l is a deformability coefficient of the rock, and p is the pressure.

In [255] is started from the time-average equation, in which the wave delay times of a 2-phase system are added up,

$$(l/v_p) = (\Phi/v_P) + [(1 - \Phi)/v_F]\,, \tag{4}$$

where Φ denotes the porosity, v_P — the wave velocity in the pore space, and v_F — the wave velocity in the solid phase. For rocks saturated with liquid this equation is quite suitable, and in the opinion of several authors it is likewise valid for gas-filled rocks having porosities of less than 20%. Certainly it is valid for eruptive rocks with very low porosities (less than 1%). Under natural conditions, porous rocks saturated with gas or liquid may change their pore volume under the influence of the external pressure. For the dependence of the porosity on the pressure applied [256], excists finds an exponential statement, which reads in a simplified form,

$$\Phi = \Phi_0\,\mathrm{e}^{-AP} \tag{5}$$

with A as an exponential factor of compaction.

The insertion of (5) into (4) leads to an equation in which, due to the similarity of the exponential dependences of K_l and Φ, a factor K_0 can be used, in such a way that in view of $K_0 = (v_F/v_{p0}) - 1$ a formula similar to equation (3) is obtained,

$$v_p/v_F = (1 + K_0\,\mathrm{e}^{-AP})^{-1}\,. \tag{6}$$

v_{p0} is the wave velocity v_p for $P = 0$.

Since the terms $K_{0l}\,\mathrm{e}^{-A_l P}$ and $K_0\,\mathrm{e}^{-AP}$ are <1, by means of series expansions of the equation (3) and (6) it can be shown that these series coincide in their first terms [257]. Fig. 2 shows for both equations the ratio of wave velocities, v_p/v_F as a function of K_{0l} and

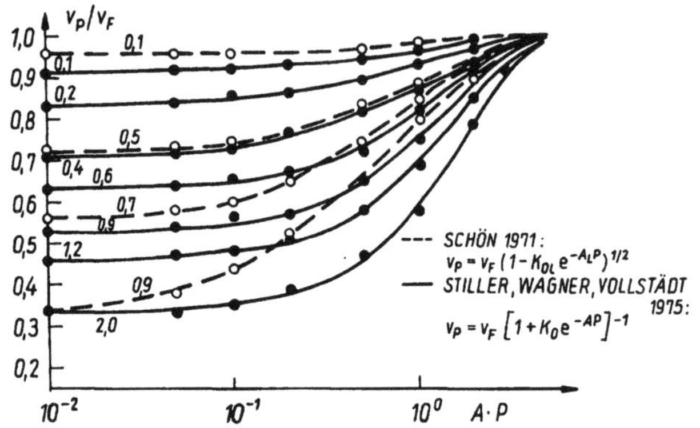

Fig. 2. Wave velocity ratio v_p/v_F as a function of K_0 (______) and K_0 (— — —) as well as of AP and A_lP.

K_0, respectively, A, P and AP. From the series expansion it follows that $A_l \approx A$. Subject to the assumptions required for the series expansions, viz., that the values of K_{0l} and K_0, respectively, are small and the values of AP are greater, these curves approach each other.

These formulae represent approximate expressions of two equations which were applicable to lunar and terrestrial [254] fractured rocks as well as to terrestrial fractured and porous rocks [258].

Qualitative estimations of K_0 and A

Subject to the assumption that at high pressures (0.5—1.0 GPa) the wave velocity v_p is very close to the wave velocity in the solid phase, v_F, K_0 can be estimated from $K_0 = (v_F/v_{p0}) - 1$.

According to our own and other authors' investigations of the wave velocity *vs.* pressure dependence [259] using v_{p0} and v_F ($v_F \approx v_p$ at high pressure) one obtains values of K_0 (Table 1).

K_0 lies between 0.04 and 1.8. Eruptive rocks have how, sedimentary rocks have high K_0 values.

The log-log representation of v_p/v_F of the equations (3) and (6) enables these curves to be used plotted wave velocity *vs.* pressure curves of lunar and terrestrial samples, with corresponding coordinate axis being parallel to each other. In this way it is possible to determine K_{0l} and K_0, respectively, as well as AP, and A_lP. Fig. 3 shows comparisons of this type for lunar and terrestrial rock samples; Fig. 3a shows them for K_{0l}, Figs. 3b—3d, for K_0. If P is known, the value of A can be determined from AP. It is also possible to determine A from deformation measurements of rocks subjected to a uniaxial pressure [260].

The ranges of A and K_0 are shown in Table 2 (1 MPa = 10 bar). The latter assumption follows from the comparison of sediments with eruptive rocks. If the transition from sediments to eruptive rocks of the earth's surface results in a decrease by about a factor of ten, it can be supposed that the transition to eruptive rocks at a greater depth likewise results in a further decrease by at least a factor of ten. K_0 must also be smaller, because due to its higher plasticity the material is likely to show a smaller number of incomplete bonds.

Table 1. v_F-, v_p-, K_0- and Φ-values for various terrestrial and lunar rocks

	v_F (km/sec)	v_{p0} (km/sec)	K_0	Φ (%)
Terrestrial rocks				
Hornfels 29a	5800	5000	0.16	
Granite 11a	6500	5900	0.10	
Hornfels T 1070	5500	5200	0.06	
Anhydrite 39a	6000	5700	0.05	
Sandstone 41 c	4500	3000	0.50	
Sandstone 44b	4500	2800	0.60	
Diabase 887	6970	6650	0.05	
Gabbro-diabase 895	6670	6500	0.06	0.2
Gabbro-diabase 898	6770	6330	0.05	
Gabbro 911	6800	6150	0.11	0.1
Gabbro 916	7240	6600	0.10	0.5
Serpentinite 928a	6500	6200	0.05	
Peridotite 925	7100	6790	0.04	
Clay 827	3250	1370	1.40	
Clay 832	3740	2390	0.57	21.9
Sandstone 690	3500	1300	1.7	15.0
Sandstone 660	5090	3870	0.42	30.0
Limestone 696	4500	3750	0.26	
Limestone 695	5350	4310	0.24	9.0
Dunite	7600	6700	0.13	1.7
Lunar rocks				
Basalt	7000	3500	1.0	
Basalt	6800	3000	1.26	
Breccia	4200	1500	1.8	

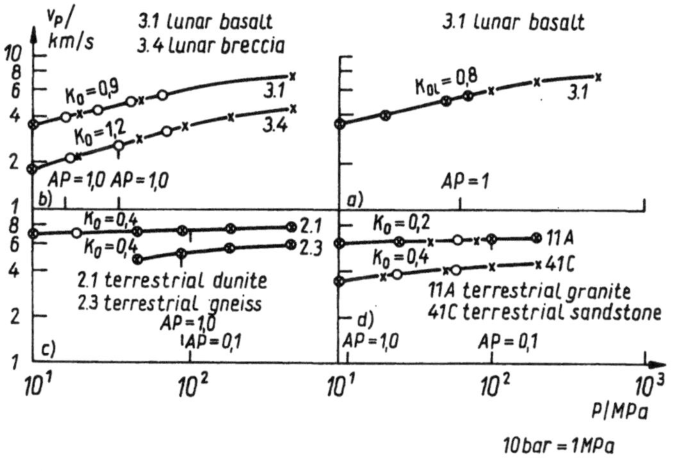

Fig. 3. Wave velocities as a function of pressure (———), and model reference curves (○○○)

Table 2. Ranges of A and K_0 for various rocks.

	A MPa^{-1}	K_0
Sediments	$10-100 \times 10^{-3}$	up to 1.8
Terrestrial eruptive rocks	$1-10 \times 10^{-3}$	up to 0.4
Terrestrial eruptive rocks at great depths (assumption)	$0.3-1.0 \times 10^{-3}$	up to 0.2

Effects of A and K_0 on the variations of wave velocity

Fig. 4 shows K_0 as a function of v_p/v_F and AP. For a given AP, i.e., for a certain material at a certain depth, the wave velocity v_p decrease with increasing K_0 and again in creases with decreasing K_0. These variations of K_0 are related to the tectonic processes before the fracture and to the fracturing phenomena in rocks investigated in the laboratory. Thus it is possible to attribute the wave-velocity variations to a parameter describing the fracturing process.

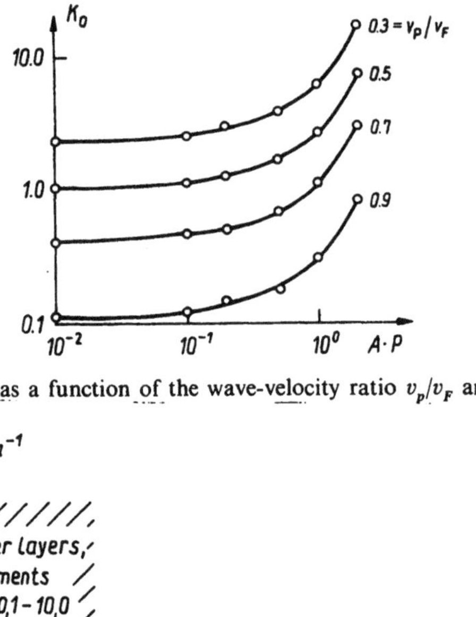

Fig. 4. K_0 as a function of the wave-velocity ratio v_p/v_F and of AP.

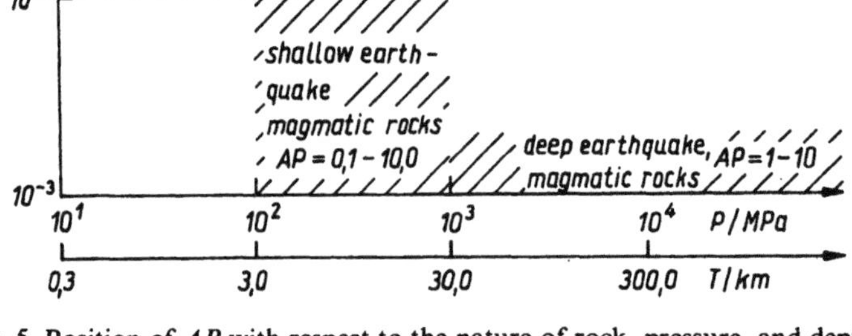

Fig. 5. Position of AP with respect to the nature of rock, pressure, and depth.

The material-depth parameter may take equal values in spite of different materials and depths.

Fig. 5 shows an AP distribution. AP values ranging from 0.1 to 10.0 may apply to: surface formations with sediments for high A and low P values, eruptive rocks in the region of shallow earthquakes for low A and higher P values, and eruptive rocks in the range of medium-depth earthquakes for low A and high P values. The latter earthquakes represent the limit of the model.

From equation (6), for $K_0\,e^{-AP} < 1$ it is possible to derive the approximate relation

$$\Delta(v_p/v_F) \doteq -e^{-AP}\Delta K_0 \,. \tag{7}$$

The function $\Delta(v_p/v_F)$ of AP and ΔK_0 is shown on the right of Fig. 6a. On the left several acid to ultrabasic rocks are plotted. Their v_F wave velocities are again set equal to the wave velocities at high pressures and taken from curves of this type. For

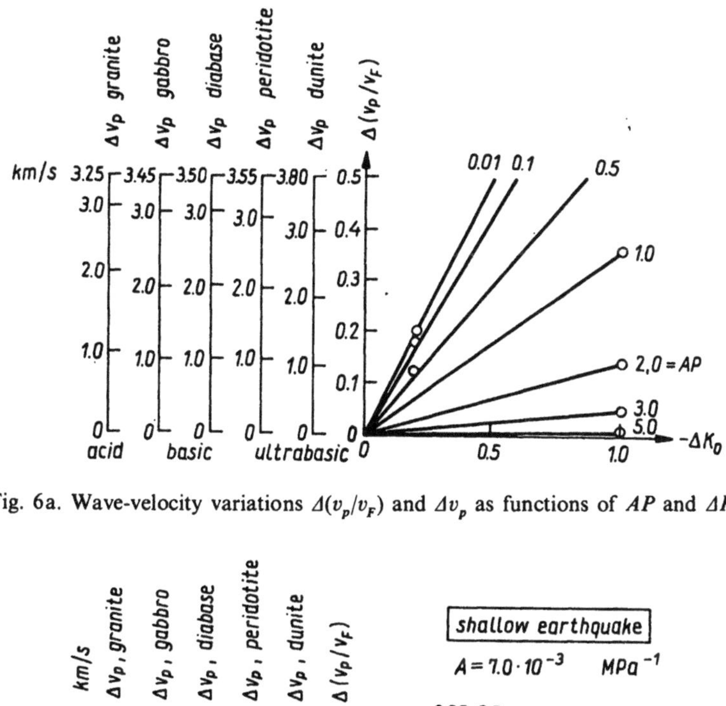

Fig. 6a. Wave-velocity variations $\Delta(v_p/v_F)$ and Δv_p as functions of AP and ΔK_0.

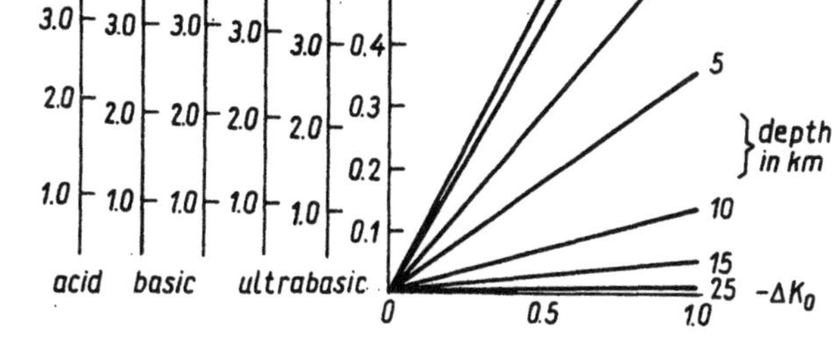

Fig. 6b. Wave-velocity variations $\Delta(v_p/v_j)$ and ΔK_0 as functions of depth and of ΔK_0.

absolute wave-velocity variations Δv_p of 1—2 km/sec, values of K_0 ranging from 0.1 to 1.0 are possible, depending on the *AP* values. Fig. 6b shows the same wave-velocity variations *vs.* the depth and ΔK_0 for shallow earthquakes.

The relationship of K_0 with the number and mean length of cracks

The relation

$$K_0 = (v_F/v_{p0}) - 1 \tag{8}$$

implies the dependence of K_0 on the wave velocity v_{p0}, which can be determined taking into account the effect of cracks on the elastic moduli for closed and open cracks [261].

Closed cracks

The wave velocity of a macroscopically homogeneous, in finitely extending medium is

$$v_{p0} = \sqrt{\frac{E_{\text{eff}}(1 - \nu_{\text{eff}})}{\varrho(1 - 2\nu_{\text{eff}})(1 + \nu_{\text{eff}})}}, \tag{9}$$

where ϱ denotes the density, E_{eff} and ν_{eff} — the effective elastic modulus and the effective Poisson ratio, respectively, of fractured rocks. For closed cracks viz., cracks with friction between the fracture faces, according to [251, 252] the following relations hold for E_{eff} and ν_{eff} (plane stress):

$$E_{\text{eff}} = E\left[1 + \frac{4\pi\bar{c}^3}{15\bar{v}}\left\{\frac{2 + 3\mu^2 + 2\mu^4}{(1 + \mu^2)^{3/2}} - 2\mu\right\}\right]^{-1} \tag{10}$$

and

$$\nu_{\text{eff}} = \nu + \frac{2\pi(1 - 2\nu)(1 - \nu^2)}{15\bar{v}}\,\bar{c}^3\left[\frac{2 + 3\mu^2 + 2\mu^4}{(1 + \mu^2)^{3/2}} - 2\mu\right]. \tag{11}$$

μ denotes the friction coefficient, for rocks ranging from 0.6 to 0.9; E and ν are the elastic modulus and the Poisson ratio, respectively, of the non-fractures rock; $\bar{c}$ is the mean crack length, and $\bar{v}$ is the volume enclosing a crack of mean crack length $\bar{c}$.

Furthermore, one has $\bar{c}^3/\bar{v} = N\bar{c}^3/V$, where V is the total volume and N is the number of cracks of mean crack length $\bar{c}$ contained in the volume V.

Let a crack distribution be assumed, where the total crack surface remains constant, so that with increasing crack length $\bar{c}$ the number of cracks N in the volume decreases; i.e., $\bar{c} = \bar{c}(N)$. Accordingly also $\bar{v}$ varies with $\bar{c}$. Fig. 7 shows the relationship between $\bar{c}$ and $\bar{v}$. If a cubical volume V of unit edge length 1 is given (central figure), then for $\bar{c} = 1$ a continuous crack surface exists and $\bar{v}$ is equal to 1. If $\bar{c}$ varies, then roughly the following values for $\bar{v}$ and $\bar{c}^3/\bar{v}$ will result with respect to this unit volume (Table 3).

According to the equation (10) and (11), E_{eff} decreases with increasing $\bar{c}$ and ν_{eff} increases with increasing $\bar{c}$. Consequently, insertion of (10) and (11) into (9) yields a depend-
a maximum. Fig. 8 shows the dependence of K_0 on $\bar{c}(N)$ for $\nu = 0.10$ and $\nu = 0.25$,
ence of v_{p0} on $\bar{c}$ in form of a minimum, while (8) yields a dependence of K_0 on $\bar{c}$ in form of

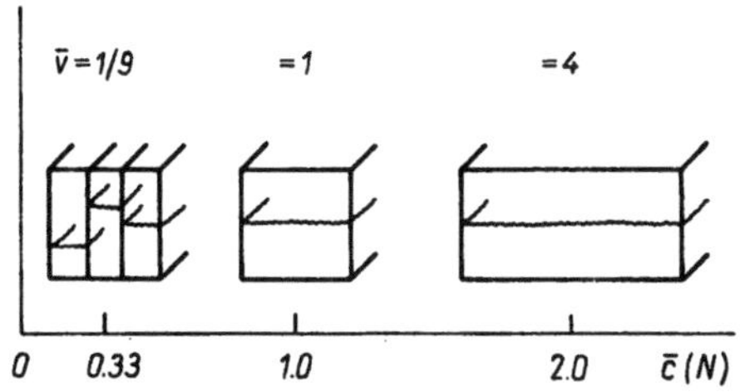

Fig. 7. The relationship between $\bar{c}(N)$ and $\bar{v}$ in unit lengths.

Table 3. $\bar{c}$, $\bar{v}$ and $\bar{c}^3/\bar{v}$ for a constant crack surface.

$\bar{c}$	$\bar{v}$	$\bar{c}^3/\bar{v}$
2	4	2
1	1	1
0.5	0.25	0.5
0.33	0.11	0.33 etc.

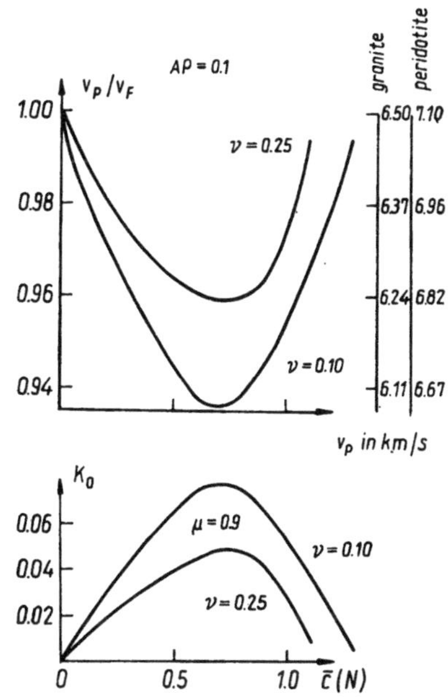

Fig. 8. K_0, v_p/v_F, and v_p as functions of $\bar{c}(N)$ for closed cracks.

$\mu = 0.9$. For the maximum of K_0,$\bar{c}(N)$ is not greater than 0.7; K_0 ranging up to 0.077. With the use of equation (6), v_p/v_F values ranging up to 0.936 are derived from this, i.e., variations in wave velocity ranging up to 6%. For granite or peridotite, these values correspond to wave-velocity variations of approximately 0.4—0.5 km/sec.

Such wave-velocity variations cannot adequately explain the variations of wave velocity occurring before earthquakes. It must be supposed that in addition to the effect of closed cracks also a certain influence is exerted by open cracks.

Open cracks

According to [250—252] the following relations hold for the elastic moduli of rocks containing open cracks.

$$\frac{1}{E_{\text{eff}}} = \frac{1}{E}\left(1 + \frac{4\pi\bar{c}^3}{3\bar{v}}\right) \text{ (plane stress),} \tag{12}$$

$$\frac{1}{E_{\text{eff}}} = \frac{1}{E}\left[1 + \frac{4\pi}{3}(1 - 2v^2)\frac{\bar{c}^3}{\bar{v}}\right] \text{ (plane strain),} \tag{13}$$

$$\nu_{\text{eff}} = \nu\left(1 + \frac{4\pi\bar{c}^3}{3\bar{v}}\right)^{-1} \text{ (plane stress),} \tag{14}$$

$$\nu_{\text{eff}} = \nu\left[1 + \frac{4\pi}{3}(1 - 2v^2)\frac{\bar{c}^3}{\bar{v}}\right]^{-1} \text{ (plane strain).} \tag{15}$$

The elastic moduli calculated with the assumption of plane stresses or plane strains show only minor differences for $\nu = 0.25$; so the equations for plane stresses are used below.

K_0 is derived using (12), (14), (9), and (8). Fig. 9 shows the dependence of K_0 on $\bar{c}(N)$ for open cracks. K_0 increases almost linearly with $\bar{c}(N)$; for $\nu = 0.25$ using $\bar{c} = 0.3$, one obtains $K_0 = 0.63$. For $AP = 0.1$ and $\bar{c}(N) = 0.3$, the ratio v_p/v_F reaches a value of about 0.6; for granite or peridotite one obtains wave-velocity variations of 2.6 and 3.8 km/sec, respectively. Although these values account for the magnitude of the occurring wave-velocity variations, in this case an explanation of the increase in wave velocity after passing through the minimum is lacking.

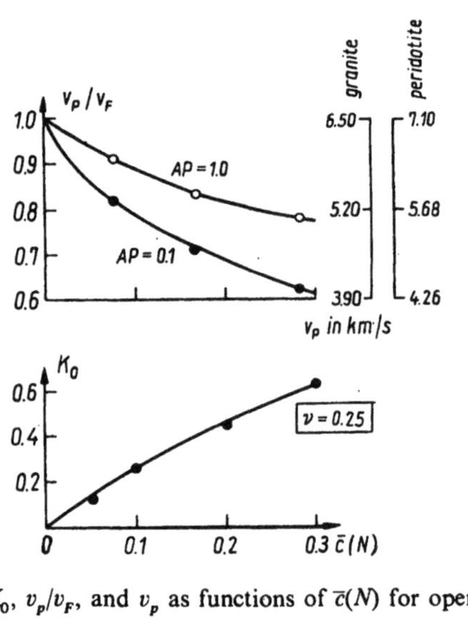

Fig. 9. K_0, v_p/v_F, and v_p as functions of $\bar{c}(N)$ for open cracks.

Probability distribution of cracks

For explaning the minimum of wave velocity for open cracks, one may attempt to start from a probability distribution of cracks. Two possibilities shall be studied: (1) an empirical probability distribution of cracks, which can be compared to the stochastic process of birth and death events, and (2) a Poisson process.

Empirical distribution: in a cubical volume $V = 1$, let the number of cracks decrease from an initial number N_0 as a function of a parameter t, which may represent the time or the tectonic stress, and let the length $\bar{c}$ increase up to a critical rupture lenght $\bar{c}_B$. We assume a distribution

$$N = N_0\, e^{-nt}\,, \qquad (16)$$
$$\bar{c}^3 = \bar{c}_B^3(1 - e^{-ct})\,, \qquad (17)$$

so that

$$N\bar{c}^3 = N_0\bar{c}_B^3\, e^{-nt}(1 - e^{-ct})\,. \qquad (18)$$

The decrease in the number of cracks and the increase in the mean crack length can be considered to be caused by the disappearance of small cracks and the development of large cracks, so that a process of this type can be compared to the birth and death processes which are well known from the theory of probability. The probability $c_j(t)$ of such a stochastic process is given by [262]

$$c_j(t) = \exp\left[-\frac{\lambda}{\mu}(1 - e^{-\mu t})\right]\frac{[\lambda/\mu(1 - e^{-\mu t})]^j}{j!} \qquad j = 0, 1, 2, 3 \qquad (19)$$

(λ, μ are parameters, $c_j(t)$ is the probability of the state j), so that by comparison with equation (19) the exponential part of equation (18) can be considered a probability distribution and $N_0\bar{c}_B^3$ can be considered a random variable. Using $n = c = l$ and the equations (18), (14), (12), (9)] and (8) one obtains the dependence of K_0 on t, ν, N_0, and $\bar{c}_B$ (Fig. 10). For

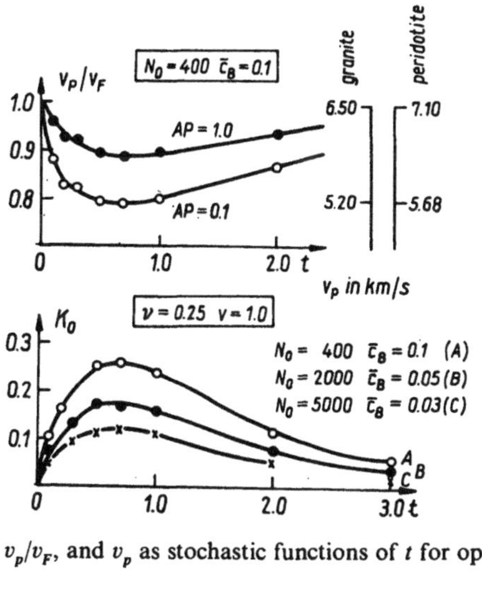

Fig. 10. K_0, v_p/v_F, and v_p as stochastic functions of t for open cracks.

$V = 1$, $\nu = 0.25$, N_0 ranging from 400 to 5000, and $\bar{c}_B$ ranging from 0.03 to 0.1, one obtains maximum values of 0.1—0.3 for K_0. Using equation (6), $N_0 = 400$, and $\bar{c}_B = 0.1$ one obtains wave-velocity variations ranging from 0.6 to 1.4 km/sec. These values correspond to the seismic wave-velocity variations with respect to both their functional behaviour and their magnitudes.

Poisson process: the representation of the wave-velocity variations as an empirical probability distribution, which was derived by comparison with the stochastic process, can be realized also under slightly modified initial conditions by means of the Poisson process.

Let the fracturing process begin with an initial crack length $\bar{c}_a(t \approx 0)$, so that the mean crack length $\bar{c}$ at the time t is given by

$$\bar{c}^3 = \bar{c}_a^3 \lambda t . \tag{20}$$

λ denotes an arbitrary parameter. According to eq. (16), let the number of cracks N at the time t be

$$N = N_0 \, e^{-\lambda t} , \tag{21}$$

so that

$$N\bar{c}^3 = N_0 \bar{c}^3 \lambda t \, e^{-\lambda t} = N_0 \bar{c}_a^3 P(X_t = i) \tag{22}$$

(i denotes the number of possible states, in this case $i = 1$). Let $P(X_t = i)$ denote the stochastic Poisson process [262]

$$P(X_t = i) = \frac{(\lambda t)^i}{i!} \, e^{-\lambda t} , \qquad i = 0, 1, 2, \dots \tag{23}$$

For $i = 1$, only two random states exist, namely $X_0 = 0$ and $X_1 = N_0\bar{c}_a^3$, so that the distribution of $N\bar{c}^3$ can be derived from a given random initial state. With $\lambda = 1$ and using equations (22), (14), (12), (9), (8), and (6) one obtains the dependence of K_0 and v_p/v_F on

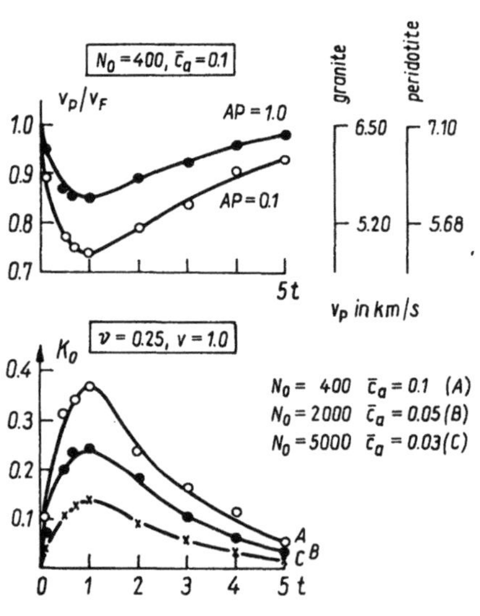

Fig. 11. K_0, v_p/v_F, and v_p as Poisson processes of t for open cracks.

t, v, N_0, and $\bar{c}_a$ (Fig. 11). With the stated values, similar wave-velocity variations are derived from Fig. 10, which likewise correspond to the seismic wave-velocity variations with respect to both their functional behaviour and their magnitude.

If it assumed that the Poisson process is applicable to fracturing, then the assumptions required for a Poisson process must likewise be valid for fracturing and it is possible to derive behaviour for fracturing processes. Such assumptions are

1. Independent increments. The increment of events, in this case of new cracks must be independent. The formation of a crack must be independent of the existence of other cracks. This is the case especially at the beginning of craching, when the developing micro-cracks are uniformly distributed over larger regions.
2. Homogeneous increments. During equal time intervals a certain well-defined number of events (cracks) occur. Possibly this means that under equivalent conditions (the same material, equal stresses etc.) the cracks formations occur again in a similar way.
3. The events occur suddenly and separately in short time intervals. This means that the crack formation manifests itself as a sudden bursting of cracks, the growth of cracks being the sudden and final expansion of the cracks of the sudden coalescence of two small cracks. However, the latter conclusions are somewhat in contradiction to assumption 1), because in this case the formation of a large crack depends on the existence of small cracks.

In addition to interpreting the wave-velocity variations, the application of known stochastic processes also permits conclusions to be drawn respect to the fracturing processes.

Effect of pore-pressure variations on the wave velocity

It is considered rather improbable that liquid movements or redistributions are possible in a crystalline rock exposed to a pressure of more than 100 MPa. Nevertheless, assuming a porous rock filled with liquid, the effect of pore-pressure variations according to equation (6) shall be investigated.

Differentiation of equation (6) with respect to P yields the equation

$$\frac{\mathrm{d}v_p}{v_p} = \frac{v_p}{v_F} K_0 A \, \mathrm{e}^{-AP} \, \mathrm{d}P \,, \tag{24}$$

which means that a variation of pore pressure, ΔP_i, produces a variation ΔP, and hence a variation $\Delta v_p/v_p$,

$$\frac{\Delta v_p}{v_p} = \frac{v_p}{v_F} K_0 A \, \mathrm{e}^{-AP} \, \Delta P \,. \tag{25}$$

The pressure determining the elastic behaviour of porous rocks consists of the external pressure P_a acting on the rock grain and the internal pore pressure P_i. While under natural stratigraphic conditions the external pressure P_a is determined by the weight of the overlying rock masses, the pore pressure results from the pressure of the gas or liquid, which in most cases corresponds to the weight of the overlying liquid column. For a first approximation the elastic behaviour of porous rocks is determined by the differential pressure $P_\Delta = P_a - P_i$. At low pressures this formula is inadequate and the rock behaviour is better described by the effective pressure $P_e = P_a - nP_i$, where the possible values of the pore-pressure modulus n are $n \leqq 1$, depending on the geological conditions [263]. For permeable reservoir rocks, $n = 0.85$. Thus the effective pressure can varied between P_Δ and P_a, depending on the geological conditions. This phenomenon, which is called the low-pressure phenomenon, is not restricted to shallow depths, but may occur in deeper regions, for which the pore pressure is high enough.

With the assumption of mineralized waters having maximum densities of 1.3 g/cm^3 and mean densities of 1.15 g/cm^3 as well as of a mean rock density of 2.5 g/cm^3, the determination of the overlying rock load and the liquid load yields

$$P = P_{\text{eff}} = P_a - nP_i = P_a - 0.4P_a = 0.6P_a, \quad P_i/P = 0.77 . \tag{26}$$

From $P = P_{\text{eff}} = P_a - nP_i$, one obtains

$$\Delta P/P = -n\Delta P_i/P \quad \text{or} \quad \Delta P/P = -0.85\Delta P_i/P . \tag{27}$$

Setting approximately $v_p/v_F \approx 1$ (if $v_p/v_F < 1$, then the values $\Delta v_p/v_p$ stated in Figs. 12 and 13 are additionally decreased), one obtains from (25):

$$\Delta v_p/v_p = 0.85K_0AP\,\mathrm{e}^{-AP}(-\Delta P_i/P) . \tag{28}$$

If, on the other hand, the tectonic pressure P_t changes, then according to (25) one has

$$\Delta v_p/v_p = K_0AP\,\mathrm{e}^{-AP}\Delta Pt/P . \tag{29}$$

The maximum of the functions (28) and (29) lies at $AP = 1$. Pore-pressure variations or tectonic pressure variations produce maximum variations of wave velocity in the region of $P = A^{-1}$. With the above A-values on obtains the following pressure and depth ranges (Z):

Sediments: $A = 10 - 100 \times 10^{-3}$ MPa^{-1}, $\quad P = 10 - 100$ MPa
$\quad Z = 0.3 - 3.0$ km

Eruptive rock: $A = 1 - 10 \times 10^{-3}$ MPa^{-1}, $\quad P = 100 - 1000$ MPa
$\quad Z = 3$—30 km.

Fig. 12 shows maximum variations of the wave velocity, $\Delta v_p/v_p$, in % as a function of K_0 and $AP = 1$, corresponding to the pressure variations $-\Delta P_i/P$ or $\Delta P_t/P$. Favourable conditions for $\Delta v_p/v_p > 2\%$ require the conditions $AP = 1$, $K_0 > 0.2$, and $\Delta P_i/P$ or $\Delta P_t/P > 20\%$

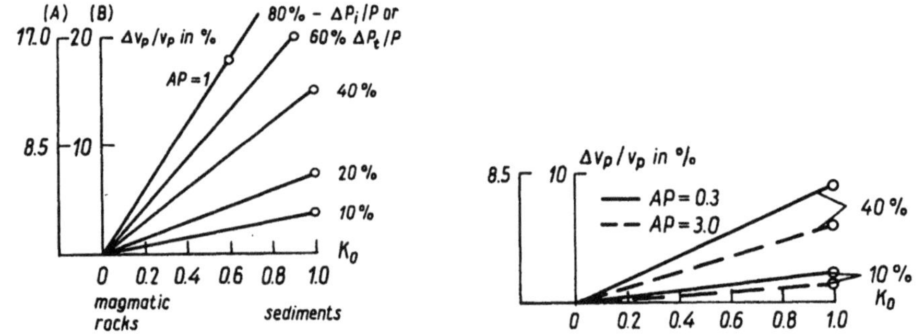

Fig. 12a. Maximum variations $\Delta v_p/v_p$ in % as a function of K_0 for variations of the pore pressure $-\Delta P_i/P$ and of the tectonic pressure $\Delta P_t/P$.

Fig. 12b. Variations $\Delta v_p/v_p$ in % as a function of K_0 for variations of the pore pressure $-\Delta P_i/P(A)$ and the tectonic pressure $\Delta P_t/P(B)$.

to be satisfied. This means a special depth, marked liquid redistributions and a considerable fractured condition. For values differring from those stated above, e.g. $AP = 0.3$ or 3.0, $K_0 \approx 0.2$, and pressure variations by 10%, the wave-velocity variations are negligible (Fig. 12). Fig. 13 shows maximum variations of wave velocity, $\Delta v_p/v_p$, in % as a function of K_0, $\bar{c}(N)$, and for pressure variations $-\Delta P_i/P$ or $\Delta P_t/P$. For pore-pressure variations of <20% and $\bar{c}(N) < 0.1$, the wave-velocity variations due to pore-pressure fluctuations are negligible. Considerable wave-velocity variations due to pore pressure fluctuations are possible only under special favourable conditions; generally they are negligible.

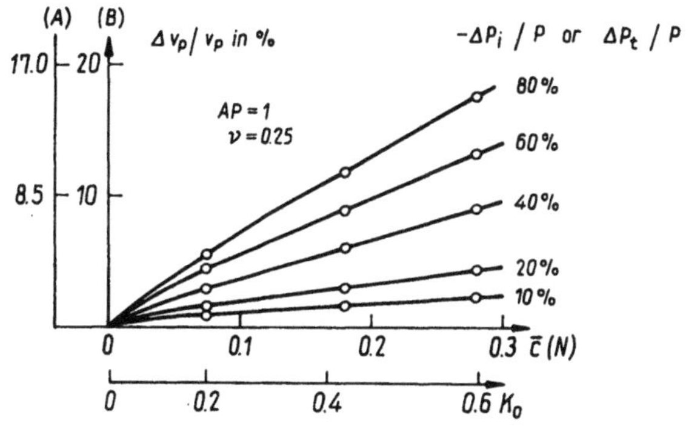

Fig. 13. Maximum variations $\Delta v_p/v_p$ in % as a function of K_0 and $\bar{c}(N)$ for variations of the pore pressure $-\Delta P_i/P(A)$ and the tectonic pressure $\Delta P_t/P(B)$.

v_s-wave-velocity variations

Equation (6) can also be applied to the wave-velocity *vs.* pressure results of v_s-wave velocities. Figs. 14 and 15 show several wave velocity *vs.* pressure curves [80, 169, 171] of several lunar and terrestrial loose rocks as compared to model curves. The Fairfax diabases were thermally pretreated in order to achieve an artificial loosening of the rock structure. They were heated up to 300 °C and 1200 °C, respectively, and then cooled down. The artificial loosening of structure distinctly manifests itself in the increase in the K_0 value from 0.1 to 1.2.

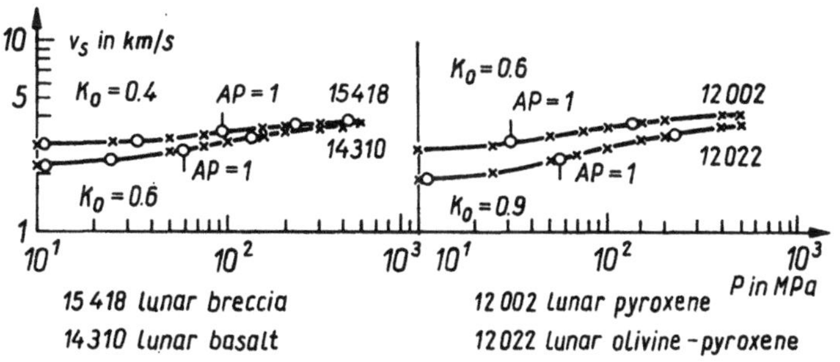

Fig. 14. v_s wave velocities as a function of pressure(———) and of model reference curves (OOO); lunar rocks.

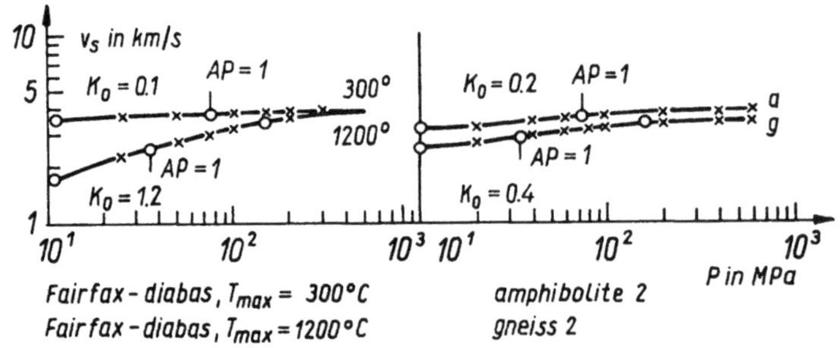

Fig. 15. v_s wave velocities as a function of pressure (———) and of model reference curves (○○○); terrestrial rocks.

Table 4. A and K_0 values for v_s-wave velocities.

	A (MPa^{-1})	K_0
Lunar breccia 155418	1.1×10^{-2}	0.4
Lunar basalt 14310	1.7×10^{-2}	0.6
Lunar pyroxene 12002	3.3×10^{-2}	0.6
Lunar olivine-pyroxene 12022	1.8×10^{-2}	0.9
Fairfax diabase 300 °C	1.3×10^{-2}	0.1
Fairfax diabase 1200 °C	2.9×10^{-2}	1.2
Amphibolite 2	1.3×10^{-2}	0.2
Gneiss 2	3.0×10^{-2}	0.4

Table 4 shows the A and K_0 values as determined from Figs. 14 and 15.

The K_0 values range up to 1.2. The A values are within a relatively narrow range from 1 to 3.3×10^{-2} MPa^{-1}.

The ratio v_p/v_s

The ratio v_p/v_s can be described by means of equation (6), if a distinction is drawn between A_p and A_s as well as between K_{0p} and K_{0s}. Then

$$\frac{v_p}{v_s} = \frac{v_{Fp}(1 + K_{0s}\,e^{-A_sP})}{v_{Fs}(1 + K_{0p}\,e^{-A_pP})}. \tag{30}$$

A reduction of v_p/v_s with increasing P requires that $A_s > A_p$ and $K_{0s} > K_{0p}$, and the increase of v_p/v_s requires a reverse behaviour.

The behaviour of the A and K_0 values can be illustrated by means of several lunar and terrestrial rocks [75, 170] (see also section 8). For that purpose it was again assumed that at the highest pressure value the wave velocities v_p and v_s coincide with v_F; K_{0p} as well as K_{0s} were calculated using equation (8). From the remaining wave velocity *vs.* pressure values, A_p and A_s were determined by means of curve fitting according to the least squares method. From equation (6) it follows that

$$Y_i = \ln\left[\frac{1}{K_0}\left(\frac{v_F}{v_i} - 1\right)\right] = -AP_1 \tag{31}$$

and

$$\sum_i (Y_i + AP_i)^2 = \min! \tag{32}$$

This gives

$$A = -\frac{\sum_i Y_i P_i}{\sum_i P_i^2}. \tag{33}$$

(A stands for A_p and A_s, respectively.)

Table 5 and 6 shows values of v_p/v_s, A_p, A_s, K_{0p}, and K_{0s} for various rocks. The wave-velocity ratio v_p/v_s is not always monotone increasing or decreasing with varying pressure; consequently, the above-mentioned relations for K_0 and A are not always defini-

Table 5. v_p/v_s values for different rocks dependent on pressure

Pressure (MPa)	0.1	10	25	50	75	100	150	200	300	400	500	
Fd 200/300 °C	1.69	1.70	1.72	1.73	1.74	1.76	1.76	1.77	1.77	1.79	1.80	↑
600 °C	1.53	1.56	1.63	1.69	1.73	1.75	1.77	1.78	1.79	1.79	1.80	↑
1000 °C	1.27	1.41	1.54	1.63	1.67	1.70	1.75	1.74	1.77	1.79	1.79	↑—
1200 °C	1.50	1.73	1.74	1.70	1.78	1.69	1.69	1.65	1.72	1.74	1.75	↑
15418.43	1.72	1.74	1.82	1.79	1.81	1.84	1.85	1.85	1.85	1.84	1.83	↑—
14310.72 A	1.89	1.89	1.90	1.87	1.87	1.87	1.86	1.84	1.83	1.82	1.83	↓+
Pressure (MPa)	**0.1**	**20**	**50**	**100**	**150**	**200**	**300**	**400**	**500**	**600**	**700**	
T. gray g.	1.70	1.69	1.68	1.70	1.71	1.72	1.73	1.73	1.74	1.75	1.75	↑
T. pink g.	1.88	1.88	1.88	1.89	1.90	1.90	1.91	1.93	1.93	1.93	1.94	↑

Table 6. A- and K_0-values for different rocks from v_p and v_s
Decrease of $v_p/v_s \rightarrow A_s > A_p$ and $K_{0_s} > K_{0_p}$

	Fairfax diabase					Lunar rocks		Texas grey granite	Texas pink granite
	200°	300°	600°	1000°		14310.72 A 1200°	15418.43		
A_p (10^{-2} MPa^{-1})	0.94		1.33	1.60	1.23	0.88	0.90	0.74	0.60
A_s (10^{-2} MPa^{-1})		1.20	1.44	1.49	1.39	0.90	0.82	0.79	0.65
K_{0_p}	0.17		0.65	2.40	2.34	0.69	0.39	0.18	0.10
K_{0s}		0.10	0.40	1.42	1.8	0.76	0.35	0.10	0.06
				—		+	—		

tely realized. In Table 5, those v_p/v_s values which on the average increase with increasing pressure are indicated by the symbol ↑, while the decreasing v_p/v_s values indicated by ↓. A definite agreement of decreasing v_p/v_s values with $A_s > A_p$ and $K_{0s} > K_{0p}$ is indicated by +, for increasing v_p/v_s values with $A_s < A_p$ and $K_{0s} < K_{0p}$, by —. Without symbol an indifferent behaviour is indicated.

However, for earthquake regions the predominant increase in the wave-velocity ratio v_p/v_s with pressure is not essential, since here the pressure variations are less important. The influence of K_0 is more serious. For these samples it can better be illustrated by the different degrees of fracturing of the Fairfax diabases. With increasing K_{0p} and K_{0s} values, v_p/v_s on the average decreases in each pressure range. In the comparison of pink granite with the grey granites, v_p/v_s likewise decreases with increasing K_{0p} and K_{0s}.

The magnitude of the A and K_0 value also shows the influence of the wave type. For A the differences between A_p and A_s are not considerable because A ist more a factor of deformability and less dependent on the wave type. On the other hand K_0 shows considerable differences between K_{0p} and K_{0s}, because the fractured condition shows larger differences between the wave types.

Conclusions

As it was shown, the results of petrophysical investigations can be employed in the explanation and description of the premonitory seismic wave-velocity variations. The following conclusions can be drawn:

1. Using equation (3) and expecially (6) it is possible to give a description of the seismic wave-velocity variations which agrees with laboratoy experiments and tectonic processes.
2. The parameter K_0 occurring in equation (3) and (6) depends on the fractured condition of the rock. A quantitative relationship between the variations of the $v_{\bar{p}}$ and $v_{\bar{s}}$ wave velocities and the variations of the K_0 parameter can be stated. Such a "K_0 model" is valid for earthquakes ranging from shallow to medium depths.
3. The K_0 parameter depends on the number and the mean length of closed and open cracks in the rock. Numerical comparisons with the seismic wave-velocity variations and the calculated values require the existence of closed and open cracks.
4. The minimum of seismic wave velocities can also be explained by means of a stochastic process. In this case the wave-velocity variations depend on an initial or final state of the number and length of cracks.

In this way the hitherto known model conceptions for the fracturing processes in the rock and the wave propagation in porous and fractured rock, respectively, can be combined in a model of the K_0 parameter.

THE EARTH'S CRUST SEISMIC MODELS FROM DSS AND LABORATORY MEASUREMENTS

M. P. Volarovich, E. I. Bayuk, V. P. Valus, I. N. Galkin

Vary important for constructing the physical model of the Earth and understanding the processes of the formation and development of its deep structure is a knowledge of how the velocities of seismic waves behave as a function of depth. There has lately been some progress in the problem of learning the vertical velocity distribution in the Earth, owing to the advent of the methods of computational seismology, and to devising new algorithms and programs for solving the inverse problem of seismology by means of the trial-and-error method. The method consists essentially in comparing observed wave characteristics with those computed for a number of alternative models; those models are then picked out which do not differ from the observations by more than the observational error [264].

The experience gained so far in studying the Earth's velocity distribution leads to the conclusion that the solution always turns out to be non-unique: the observations agree with a whole set of physically different models which are characterized by a certain range of variation in the velocity parameters. The non-uniqueness can be weakened by increasing the precision and detailedness of the experiment, and by making use of a large number of characteristics (both kinematic and dynamic ones) of primary and secondary waves.

The results of seismic investigations are represented by a strip in the plane "velocity-depth", a strip that symbolizes the discriminative power of the seismic method. The strip can be made thinner, and the velocity distribution be interpreted in physical terms only on the basis of a joint analysis of seismic and other geophysical data.

The present paper reports the results of comparing seismic data about the velocity distribution of the crust with direct measurements of the velocities of elastic waves in various rock samples under pressure. In contrast to similar previous studies, seismic results are used that have been obtained by an objective method, and which describe the whole range of possible solutions, together with mathematically treated results of laboratory data.

Velocity models for crusts of various types

The structure of the crust has been studied from the data supplied by seismology and deep seismic sounding by means of the trial-and-error method with the help of a computer program [265], for regions with different types of crust structure: continental mountainous type (Tyen Shan, the profile Andizhan-Dushanbe), continental transition type (the island of Sakhalin, the pass of Tatarsky), oceanic type (the Pacific). The computations have been done for a thick-layer, local-laterally homogeneous and isotropic structure of the crust according to the formulas of the null approximation in the ray method of the dynamic theory of wave propagation [266]. 600—900 theoretical models have been treated. The number of models that agree with the observations are [267—270]: 46 for Tyen Shan (the standard deviation of the P-wave travel times being 1 sec., that for PP — 1.5 sec., and that for the phase velocity of Love waves 0.05 km/sec.), 27 for Sakhalin (the accuracy of the readings of the first P being 0.2 sec., of the amplitudes — 0.2 of a logarithm unity), and 14 for the Pacific (the error being 0.2—0.3 sec. and 0.3 of a logarithm unity respectively). Those which have "passed" are shown in Fig. 1. The arabic figures indicate the uncertainty of the velocity, of its gradient, and of interface depths.

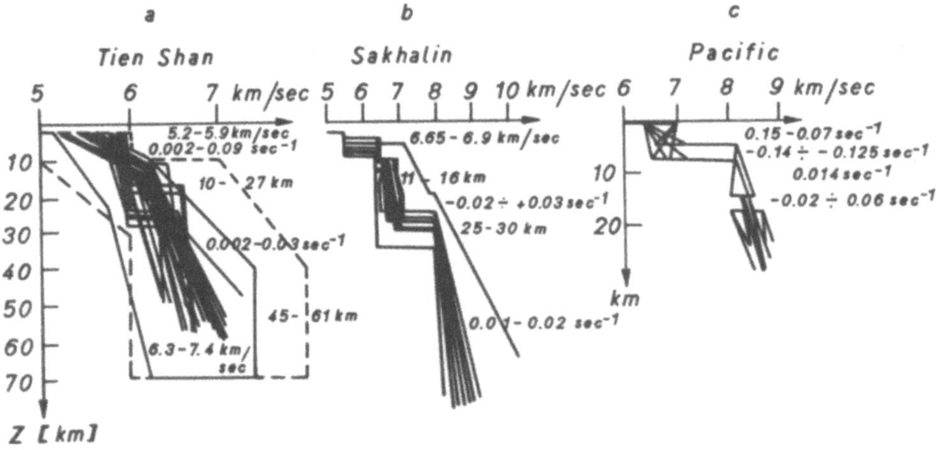

Fig. 1. Velocity cross-sections of the Earth's crust, determined by the trial-and-error method on an electronic computer for the crust of the continental mountainous type (Tyen Shan), continental transition type (Sakhalin), oceanic type (the Pacific). The Arabic figures show the uncertainty range of the parameters corresponding to the cross-sections.

Non-uniqueness in the velocity distribution has been found to exist for all the three types of crust structure. It implies an uncertainty in the parameters as well as a basic difference in the models: for Tyen Shan there are discontinuities of the first and the second kind, for the transition zone between Asia and the Pacific — a low velocity channel and a positive velocity gradient in the crust. On the whole, the velocity strips show an essential difference to exist in the three regions studied. There is a decrease in crust thickness and an increase in the velocities when going from the continent towards the ocean.

Velocities in rock samples under pressure

P-wave velocities in rock samples have been determined at hydrostatic pressures upt to 4 kb, and at quasihydrostatic ones — up to 20 kb at room temperature. We studied specimens of igneous rocks from various continental regions of the Soviet Union: the Kola peninsula, the North Caucasus, Central Kazakhstan, Sakkalin, and some others [271—275]. The experimental data on P-wave velocities obtained for more than 100 granite samples,

75 basic rocks (gabbro), and 80 ultrabasic rocks have been used to determine velocity histograms. Then the normal probability law has been fitted to the histograms, with the probability density given by

$$f(x) = \frac{1}{\sigma\sqrt{2\pi}} e^{-\frac{(x-\bar{v})^2}{2\sigma^2}}.$$

As is known, the parameters of the normal law are: $\bar{v}$ — the mean value, and σ — the standard deviation.

The probability density graphs, or velocity distributions for granites, gabbro, and ultrabasic rocks are shown in Fig. 2 at a pressure of 1 atm, as well as at those of 1, 4 and 15 kb.

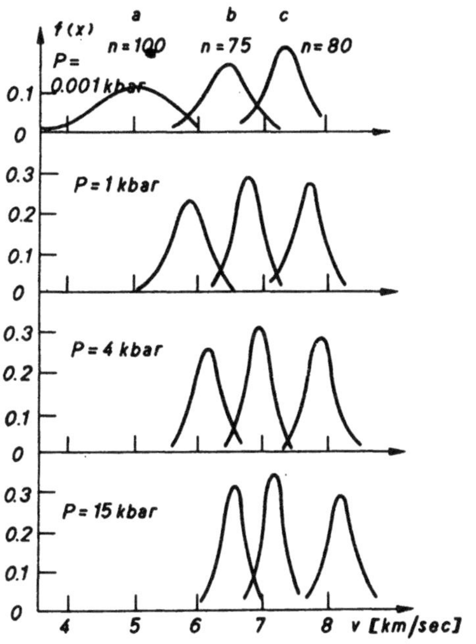

Fig. 2. Graphs of the probability density of the *P*-wave velocity for granites (a), gabbro (b) and ultrabasic rocks (c) at various pressures.

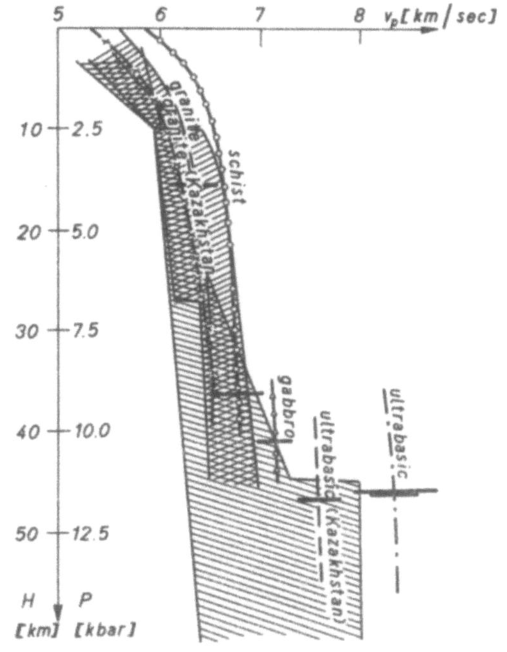

Fig. 3. The strip of seismic velocities (shaded) and the results of laboratory measurements; the vertical lines represent the mean velocities as functions of pressure (depth), the horizontal ones are strips for the crust of the continental mountainous type (Tyen Shan).

As the pressure increases, the scatter in the velocities decreases, and accordingly the standard deviation decreases. Further, the velocity distribution systematically moves towards higher velocities, since the mean value of the velocity increases. The parameters of the normal distribution law at various pressures are given in the table.

The mean velocities have been calculated also for some rock types of Central Kazakhstan [275]: granites, gabbro and schists. The data here are more scanty, so that the parameters of the normal distribution law are less accurate. The same holds for the Sakhalin basalts [274, 276].

Table 1. The parameters of the normal distributions law for igneous rocks.

		1 atm	1 kb	4 kb	15 kb
		granites			
$\bar{v}$	km/sec	5.06	5.87	6.17	6.56
σ	km/sec	0.33	0.26	0.23	0.20
		basic rocks			
$\bar{v}$	km/sec	6.42	6.78	6.98	7.17
σ	km/sec	0.35	0.20	0.18	0.17
		ultrabasic rocks			
$\bar{v}$	km/sec	7.31	7.70	7.90	8.20
σ	km/sec	0.28	0.23	0.23	0.24

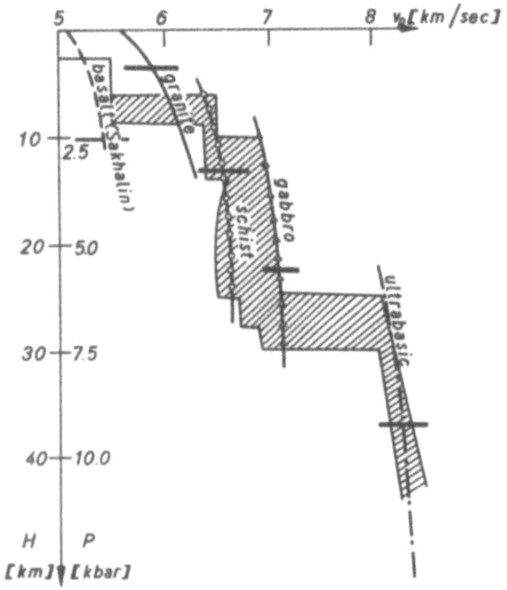

Fig. 4. The strip of seismic velocities (shaded) and the results of laboratory measurements; the vertical lines represent the mean velocities as functions of pressure (depth), the horizontal ones are strips for the crust of the continental transition type (Sakhalin).

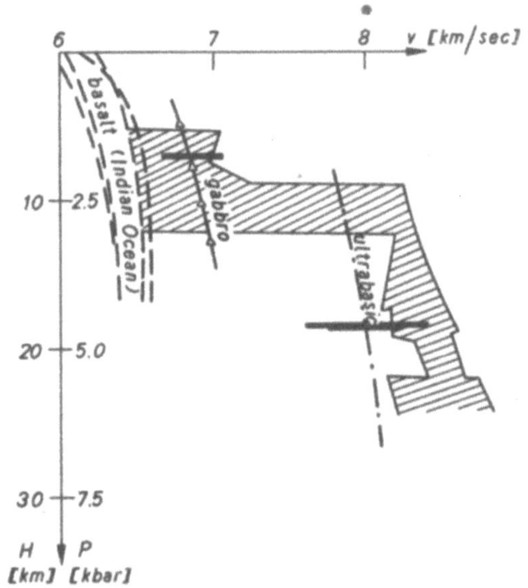

Fig. 5. The strip of seismic velocities (shaded) and the results of laboratory measurements; the vertical lines represent the mean velocities as functions of pressure (depth), the horizontal ones are strips for the oceanic crust.

For the oceanic crust we compared the data of seismology and deep seismic sounding with laboratory determinations of velocities in rocks from the Atlantic [277].

The results of the comparison are shown in Fig. 3, 4, 5. Shaded are the velocity strips from seismic data, lines with various signs — the curves of mean velocities in various samples as functions of pressure (depth), the velocity range for 70% of the samples is given in horizontal shading.

The velocity cross-section fot Tyen Shan down to depths of 30—40 km is similar to the region where the curves for granite are situated. The width of the region is such that the deviation of velocity values from the mean is equal to $\pm\sigma$. In this way an explanation can be found for the variant with no velocity discontinuity in the crust. Just above the Moho seismic velocities reach values characteristic for gabbro. It should be noted that the curves of the mean velocities as functions of pressure for granites and gabbro are almost identical for the whole set of the continental samples and the samples from Kazakhstan. They differ, however, for ultrabasic rocks (the mean velocity is lower for Kazakhstan). The seismic velocities in the upper mantle under Tyen Shan are situated in the 80% strip between the two curves of the mean velocities.

The four-layer seismic model for the consolidated crust and upper mantle under Sakhalin finds a clear lithological interpretation when we compare it with laboratory data: the velocity in the upper layer of the crust falls into the left (low-velocity) part of the 80% values of "granite" velocities and the velocities in the Sakhalin basalt. The velocities in the intermediate layer are identical with those for schists, and the velocities in the lower layer approximate the velocities for gabbro. The curve of the mean velocity for ultrabasic rocks falls into the centre of the velocity strip for the upper mantle (Fig. 4).

An analysis of the velocity models for ocean is more difficult, on account of sparsity of rock samples (especially the mantle ones). We have plotted a few curves in Fig. 5 for basalts from the rift zone of the Indian ocean. If we compare them with the data averaged over the whole Earth it is found that the oceanic crust has the velocities of gabbro, and the upper mantle falls in to the right part of the 80% strip of "ultrabasic" velocities (Fig. 5).

The general conclusion is that a joint use of seismic and laboratory data has proved to be fruitful for preliminary determinations of the constitution of the principal crust layers in regions differing as to structure. The possibility to narrow the non-uniqueness strip of seismic data with the help of laboratory measurements of velocities should be tested in concrete, localized territories.

Lastly, there is one more aspect of the comparison of independent data on the velocities of elastic waves. With measurements of rock samples under pressure, we can prognosticate seismic cross-sections, the wave-field peculiarities, and the technique of seismic observations

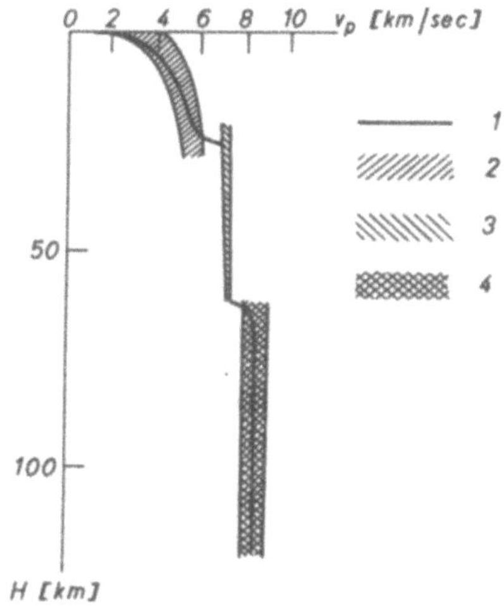

Fig. 6. Velocity as a function of depth a lunar seismograns (*1*) and for lunar samples: *2* — basalt, *3* — gabbro, *4* — ultrabasic.

in regions where no observations have yet been taken, or they are scanty. An example is provided by the experience of seismic investigations on the Moon [278].

Finally, there is one more aspect of comparing the independent data concerning the velocities of elastic waves. With the measurements of samples under pressure being known, it is possible to predict the seismic section, the wave field peculiarities and the technique of seismic observation in the regions where it has not been carried out, or where it is insufficient. The experience of seismic investigations on the Moon may serve as an example of such a case.

One of the possible versions of the velocity section of the moon crust and the mantle top down to a depth of 120 km has been plotted in accordance with the single hodograph of the first and the next arrivals of longitudinal waves within a range of epicentral distances equal to 67—356 km and 900—1100 km as a result seismic observation in the Fra Mauro region of the Ocean of Storms. Growth of velocity is observed in the crust upper portions due to consolidation of rock and closing of macrodefects produced by meteorite bombardment. A velocity jump of from 7 to 8 km/sec is notices at a depth of about 65 km. A boundary of the first kind is supposed to be in the crust at a depth of about 25 km in one of the versions. Comparison of the seismic section with the results of laboratory measurements of velocities in moon samples under respective pressures shows that the velocities in the upper part of the consolidated moon crust are close to the velocities in sea basalts, those in the crust lower parts are close to the gabbro-anorthosites of the moon mountainous regions, while in the mantle top they are close to the terranean ultrabasites (Fig. 6).

MATHEMATICAL MODEL FOR CALCULATION OF THE PROBABILITY OF SUDDEN STEPWISE PHASE TRANSFORMATION OCCURENCE

J. L. KOPYSTYŃSKI

Polymorphic phase transformations taking place in the Earth's interior, i.e. under high pressure, may be a direct or indirect cause for seismic focus formation. Therefore, they are of interest in geophysics.

The results of the studies of polymorphic phase transformations presented here are based on experimental investigations carried out in the High Pressures Laboratory of the Earth Physics Institute of the CNRS, France, [279, 280, 283], as well as on theoretical research related to these experiments [281—283].

One of the parameters that undergo a sudden change in the polymorphic transformation process is the specific volume: as pressure increases and becomes greater than the phase equilibrium pressure P_T, the medium in the phase 1 state (low pressure state) undergoes transformation into the phase 2 state (high-pressure state). The phenomenon of such a transformation corresponds to the decrease in the specific volume. In the opposite, phase 2 → phase 1 transformation, which takes place when pressure decreases and becomes lower than P_T, the specific volume increases.

The experiments on polymorphic transformations have been made for samples of alkali halides, namely RbCl or KCl, whose phase equilibrium pressure P_T are 5.2 kb or 17 kb, respectively.

The piston-cylinder experimental arrangement enabled us to make simultaneous measurements of the change of valume and the pressure within the high-pressure chamber. The sample under study was placed in a liquid medium, which provided the hydrostatic pressure. It was noted that depending upon the preparation of the sample, notably its preliminary thermal treatment, the relation between the changes in pressure and volume behaved in different manner: the hysteresis loop of the phase transformation could change its width, and the branch of the loop corresponding to the transformation from the low-pressure to high-pressure phase could be either continuous or stepwise. The latter case is related to the situation in which certain domains of the medium undergo sudden transformations. Series of few consecutive stepwise transformations (in polycrystalline samples) or individual, intense transformations followed by a continuous transformation (in monocrystals) have been observed. We have not observed stepwise transformations in the high-pressure → low pressure phase transformation.

As concerns the relationship between the polymorphic transformation and the earthquake focus generation, one can outline the following two schemes, dependent upon the type of the transformation:

1. If the transformations is continuous, the sequence of events in the neighbourhood of the volume which undergoes transformation is: increase (or decrease) of pressure in the region surrounding the transformable medium → continuous phase transformation from phase 1 to phase 2 (or from phase 2 to phase 1) → increase in stresses on the boundaries between the transformed and nontransformed regions → violation of the mechanical equilibrium → relative motion of layers, i.e. an earthquake focus formation.

2. If transformation is sudden, then the sequence is: increase (or decrease) of pressure in the region surrounding the transformable medium → creation of the metastable state → sudden transformation from phase 1 to phase 2 (or from phase 2 to phase 1) accompanied by an abrupt decrease (or increase) of the volume which is transformed, i.e. the implosion (or explosion) → earthquake focus formation due to relative motion of layers or fast displacement of the transformation front.

In what follows we will construct a mathematical model describing stepwise transformations from phase 1 to phase 2. Since such transformations have been observed, we are able to make an experimental verification of the model proposed.

The aim of this study is to describe a mathematical model permitting the calculation of the probability of new successive sudden transformation occurrence, based on the observation of a previous series of stepwise transformations. The comparison between the results calculated from the presented model and those obtained experimentally with rubidium chloride is presented. In the experiment in which the volume of the high-pressure chamber changes (decreases) at a constant rate A, the dependence between pressure P, time t and the transformation advancement function is given by

$$\frac{P - P_T}{P_T} = \frac{At - \lambda_1 v}{\lambda_2 + \lambda_3 v}. \tag{1}$$

Time is measured from the point in which pressure reached P_T; parameters λ_1, λ_2, λ_3 are experimental constants.

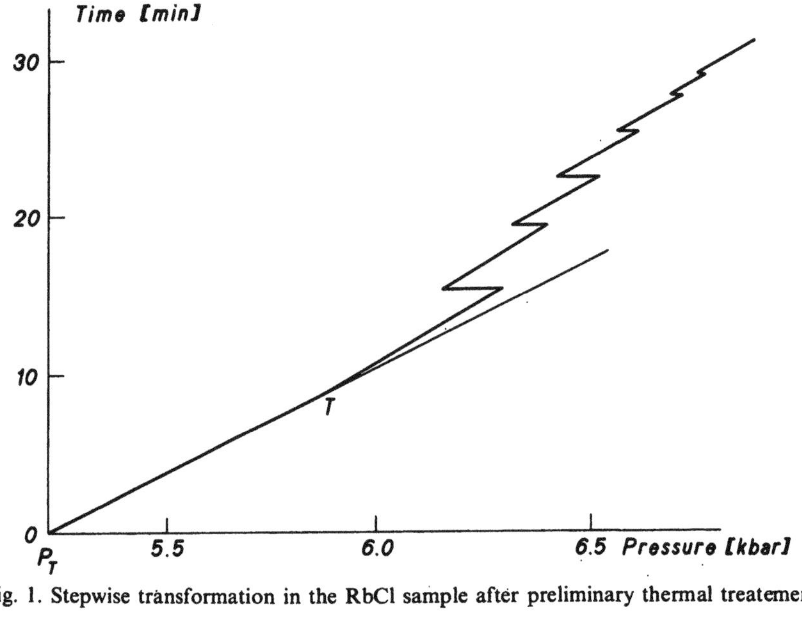

Fig. 1. Stepwise transformation in the RbCl sample after preliminary thermal treatement.

Fig. 1 shows an exemplary, experimental plot of pressure *vs.* time in the polymorphic transformation process in rubidium chloride. Segment OT corresponds to the compression of the sample; continuous transformation begins to be noticeable at point T; the horizontal segments correspond to consecutive sudden transformations.

As it can be seen in Fig. 1 on the pressure diagram, the slopes of the segments between stepwise transformations are practically identical. This means that within a measurement accuracy the total compressibility coefficient of the surrounding liquid and the sample remains constant. Thus in Fig. 1 pressure drops only correspond to the phase transformation. During each pressure drop a certain defined fraction of the sample mass is entirely transformed. Fig. 2 shows the scheme of experimental relation used to construct the mathematical model: there are only stepwise transformations, and segments OA, BC, DE, ... are parallel.

Let us assume that during the i-th step the domain of volume v_i (where v_i is the volume reduced to transformation pressure P_T) is transformed. The function of reaction advancement after the i-th partial transformation can be expressed by[1])

$$\nu = \sum_{j=1}^{j=i} \nu_j \equiv \nu_{(i)}, \tag{2}$$

where: $\nu_j = v_j/V_1$ and V_1 is the initial volume of the sample in the low pressure phase, under pressure P_T. The time dependence of the pressure between consecutive partial transformations is presented according to Fig. 2 by segments OA, BC, DE, ... , defined by the equations

$$\left.\begin{array}{lll} P = P_T(1 + ct) & \text{for} & 0 \leqq t \leqq t_1 , \\ P = P_T(1 + ct) - \Delta_1 & \text{for} & t_1 < t \leqq t_1 + \tau_1 , \\ \cdots & & \\ P = P_T(1 + ct) - \Delta_{(i)} & \text{for} & t_1 + \tau_{(i-1)} < t \leqq t_1 + \tau_{(i)} . \end{array}\right\} \tag{3}$$

Introducing the first of equations (3) in equation (1), where $\pi = 0$ since $t < t_1$, we obtain

$$c = A/\lambda_2 . \tag{4}$$

The introduction of consecutive equation (3) into equation (1) leads to the following relations describing the function of reaction advancement:

$$\nu_i = \frac{\lambda_2 \Delta_{(i)}}{P_T \lambda_1 + P_T C \lambda_3 \theta_i}, \tag{5}$$

where:

$$\theta_i = \begin{cases} t_1 - \dfrac{1}{P_T C} \Delta_1 & \text{for} \quad i = 1 , \\ t_1 + \tau_{(i-1)} - \dfrac{1}{P_T C} \Delta_{(i)} & \text{for} \quad i \geqq 2 . \end{cases} \tag{6}$$

[1]) To simplify the notation the indices in parentheses mean summing up of the partial values of a given quantity $\sum_{j=1}^{j=i} x_j \equiv x_{(i)}$.

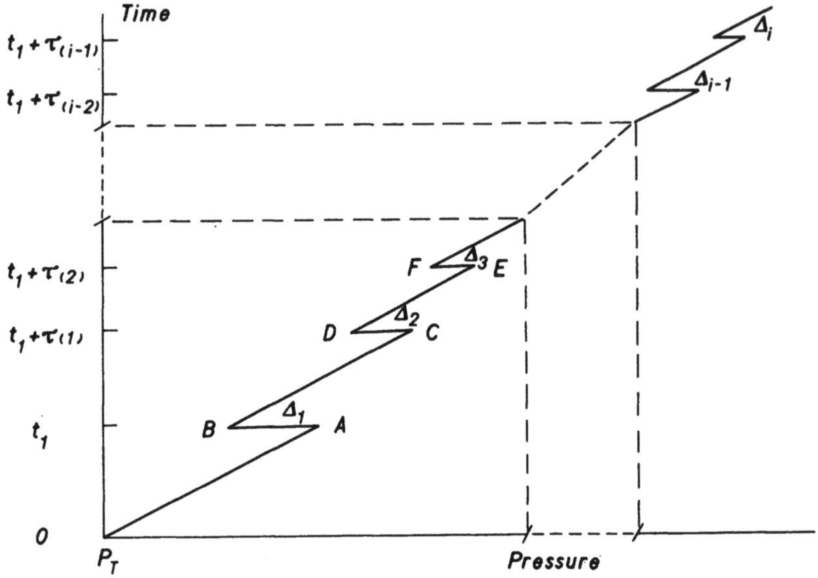

Fig. 2. Theoretical scheme of consecutive stepwise transformations (the case in which the volume of the high-pressure chamber decreases linearly with time).

From equations (5) and (2) the recurrent formula for the volume transformed during the i-th step is obtained

$$v_i = \frac{\lambda_2 V_1}{P_T} \frac{\Delta_{(i)}}{\lambda_1 + \lambda_3 C \theta_i} - v_{(i-1)} . \tag{7}$$

The probability dp_i that a certain domain of volume $v_{i,P}$ will undergo a sudden transformation in time, between τ and $\tau + d\tau$ is proportional in the first approximation to

1. the rate of potential grain activation of a new phase; this according to [281] is proportional to the factor $[(P - P_T)/P_T]^r$;
2. mass fraction $(1 - v_{(i-1)})$ of the sample, which has not been transformed as yet;
3. domain volume $v_{i,P}$ raised to the s-th power ($v_{i,P}$ denotes the volume under pressure P; the parameter s depends on the geometry of domain growth[2]);
4. time interval $d\tau$ in which the transformation is expected.

Thus the probability of the occurrence of the i-th transformation is

$$dp_i(v_{i,P}, d\tau) = B\, e^{-\zeta\tau} \left(\frac{P - P_T}{P_T}\right)^r (1 - v_{(i-1)})\, v_{i,P}^s \, d\tau . \tag{8}$$

The parameters B and ζ are positive, and time τ is measured from the beginning of the previous $(i - 1)$-th transformation. A decreasing exponential factor was introduced into formula (8) to express the experimentally known behaviour of a stepwise transformation amplitude with increasing over-pressure $(P - P_T)$.

[2]) Discussion in paper [282] shows that $s = {}^2/_3$ or $s = {}^1/_3$ if the activation of new domains of the new phase takes place on the surface of these domains or in their entire volume, respectively.

The number of free parameters in equation (8) can be reduced to three by applying the condition of probability normalization:

$$\int_0^\infty \frac{\mathrm{d}p_i}{\mathrm{d}\tau}\,\mathrm{d}\tau = 1\,. \tag{9}$$

It is obviously that condition (9) is applicable only when it is assumed in advance that the transformation considered will certainly occur in volume v_i. The applicability of this condition is therefore limited to the statistical treatment of experimental data.

According to equation (3) the factor $\frac{P - P_T}{P_T}$ is of the following form

$$\frac{P - P_T}{P_T} = \begin{cases} C\tau & \text{for} \quad i = 1\,, \\ C(\theta_{i-1} + \tau) & \text{for} \quad i \geqq 2\,. \end{cases} \tag{10}$$

Introducing relation (10) in equations (8) and (9) after integration we obtain

$$\left.\begin{aligned} p_1 &= \frac{\gamma(r + 1, \zeta\tau)}{\Gamma(r + 1)} \\ B &= \frac{\zeta^{r+1}}{C^r v_1^s}\,\frac{1}{\Gamma(r + 1)} \end{aligned}\right\} \quad \text{for} \quad i = 1\,,$$

$$\left.\begin{aligned} p_i &= \frac{\gamma[r + 1, \zeta(\theta_{i-1} + \tau)] - \gamma(r + 1, \zeta\theta_{i-1})}{\Gamma(r + 1) - \gamma(r + 1, \zeta\theta_{i-1})} \\ B &= \frac{\zeta^{r+1}}{C^r \cdot v_2^s}\,\frac{\mathrm{e}^{-\zeta\theta_{i-1}}}{1 - \nu_{(i-1)}}\,\frac{1}{\Gamma(r + 1) - \gamma(r + 1, \zeta\theta_{i-1})} \end{aligned}\right\} \quad \text{for} \quad i \geqq 2\,, \tag{11}$$

where: $\Gamma(r + 1)$ is Euler's integral, and $\gamma(r + 1, x)$ is incomplete Euler's integral,

$$\left.\begin{aligned} \Gamma(r + 1) &= \int_0^\infty y^r\,\mathrm{e}^{-y}\,\mathrm{d}y\,, \\ \gamma(r + 1, x) &= \int_0^x y^r\,\mathrm{e}^{-y}\,\mathrm{d}y\,. \end{aligned}\right\} \tag{12}$$

Finally, for i stepwise transformations observed, occurring at the ends of time intervals $t_1, \tau_1, \ldots, \tau_{i-1}$, we have $2i$ equations (11) with parameters ζ, B, r, and s. Each pair of equations corresponds to pressure $P_1, P_2, \ldots, P_i$ of the consecutive stepwise transformation.

To use equation (11) it is necessary to establish a certain value of the probability of rapid transformation occurrence, as it was already mentioned when introducing equations (8) and (9). Having the experimental data on stepwise transformation series in hand, it is possible to adopt the following consideration: the i-th transformation originating in time $t_i = t_1 + \tau_{(i-1)}$ divides two such time intervals $0 < t \leqq t_i$ and $t > t_i$ that the probability of transformation in both intervals is the same and equal to 0.5. This assumption has statistical meaning for transformation series. For a single transformation, which has already occurred in time t_i, the probability in time interval $0 < t \leqq t_i$ is of course equal to 1.

For the assumed values of probabilities p_i equations (11) form $2i$ equations with four unknown parameters ζ, B, r, and s. Experimental data from two stepwise transformations are formally sufficient to find these parameters. In practice, however, to obtain more accurate values of these parameters, a large set of statistical data, which would give the average values of the model parameters$\bar{\zeta}, \bar{B}, \bar{r}$, and$\bar{s}$ is needed.

If data related to i consecutive transformations are available, then from equation (8) the probability of the $(i + 1)$-th transformation, which may occur in a volume range of v to $v + \mathrm{d}v$ and in a time interval of τ to $\tau + \mathrm{d}\tau$, can be calculated

$$\mathrm{d}p_{i+1}(\mathrm{d}v, \mathrm{d}\tau) = B_V \bar{B}\, \mathrm{e}^{-\bar{\zeta}\tau} \left(\frac{P(t) - P_T}{P_T}\right)^{\bar{r}} (1 - v_{(i)})\, v^{\bar{s}}\, \mathrm{d}v\, \mathrm{d}\tau\,, \tag{13}$$

where: $t = t_1 + \tau_{(i-1)} + \tau$, and B_v is the normalizing constant related to volume.

Since in equation (13) the variables τ and v are separated, the normalization can be made independently for time and volume. For the $(i + 1)$-th transformation from equation (8) it follows that

$$\frac{P(t) - P_T}{P_T} = C(\theta_i + \tau)\,. \tag{14}$$

After introducing this relation into formula (13) and integration, the probability of transformation occurrence in time interval τ (measured from the beginning of the previous i-th transition) and in the domain with a volume between ε and v can be expressed:

$$p_{i+1}(v, \tau) = B_V \bar{B} C^{\bar{r}} (1 - v_{(i)}) \int_{\varepsilon}^{v} v^{\bar{s}}\, \mathrm{d}v \int_{0}^{\tau} (\theta_i + \tau)^{\bar{r}}\, \mathrm{e}^{-\bar{\zeta}\tau}\, \mathrm{d}\tau\,. \tag{15}$$

The volume ε in a physical sense corresponds to the critical volume of the grain. The normalization condition is of the form

$$p_{i+1}(V, \infty) = B_V \bar{B} C^{\bar{r}} (1 - v_{(i)}) \int_{\varepsilon}^{V} v^{\bar{s}}\, \mathrm{d}v \int_{0}^{\infty} (\theta_i + \tau)^{\bar{r}}\, \mathrm{e}^{-\bar{\zeta}\tau}\, \mathrm{d}\tau = 1\,. \tag{16}$$

This condition means that in an infinitely long time interval a stepwise transformation will occur in a certain domain with the volume comprised between ε and total transformable volume V. The total volume is

$$V = V_{1,P_T}(1 - v_{(i)})[1 - a_1(P - P_T)]\,, \tag{17}$$

where: V_{1,P_T} is the volume of the low pressure phase under transformation pressure, and a_1 is its coefficient of compressibility.

Dividing relation (15) by relation (16) the normalized formula for the probability is obtained

$$p_{i+1}(v, \tau) = \frac{\int_{\varepsilon}^{v} v^{\bar{s}}\, \mathrm{d}v}{\int_{\varepsilon}^{V} v^{\bar{s}}\, \mathrm{d}v} \cdot \frac{\int_{0}^{\tau} (\theta_i + \tau)\, \mathrm{e}^{-\bar{\zeta}\tau}\, \mathrm{d}\tau}{\int_{0}^{\infty} (\theta_i + \tau)\, \mathrm{e}^{-\bar{\zeta}\tau}\, \mathrm{d}\tau}\,. \tag{18}$$

The first factor in this formula, after integration and when $\varepsilon \to 0$, is

$$\left.\begin{array}{ll}\left(\dfrac{v}{V}\right)^{\bar{s}+1} & \text{for} \quad \bar{s} > -1\,, \\ 1 & \text{for} \quad \bar{s} \leqq -1\,. \end{array}\right\} \tag{19}$$

Further on the case when $\bar{s} > -1$ is considered. This corresponds to such a transformation whose probability depends upon the transforming domain. Integration of the second factor in formula (18) (using relation (12)) gives

$$p_{i+1}(v, \tau) = \frac{\gamma[\bar{r}+1, \bar{\zeta}(\theta_i+\tau)] - \gamma(\bar{r}+1, \zeta\theta_i)}{\Gamma(\bar{r}+1) - \gamma(\bar{r}+1, \zeta\theta_i)} \left(\frac{v}{V}\right)^{\bar{s}+1} \tag{20}$$

This equation expresses the transformation probability of a certain volume comprised between ε and $v < V$. The formula for the transformation probability of volume v_{i+1} comprised between a certain minimum volume v and the total volume V is

$$p_{i+1}(v < v_{i+1} < V, \tau) = \frac{\gamma[\bar{r}+1, \zeta(\theta_i+\tau)] - \gamma(\bar{r}+1, \zeta\theta_i)}{\Gamma(\bar{r}+1) - \gamma(\bar{r}+1, \zeta\theta_i)} \left[1 - \left(\frac{v}{V}\right)^{\bar{s}+1}\right]. \tag{21}$$

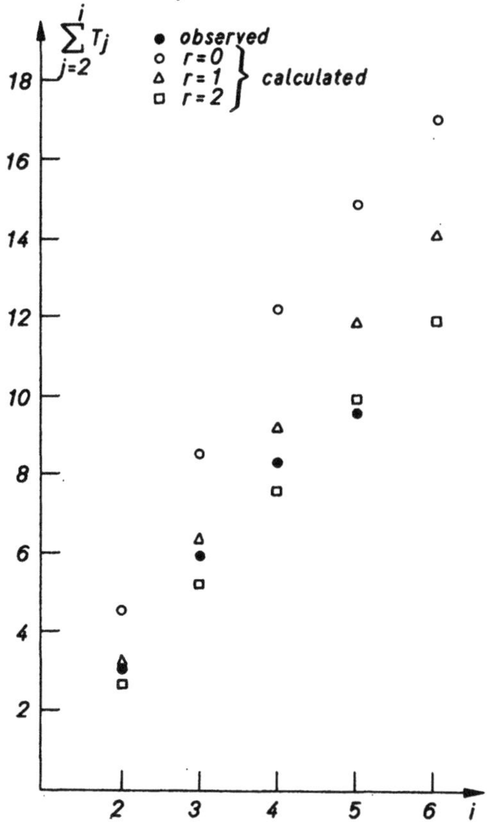

Fig. 3. The relationship between the time of occurrence and the consecutive transformation number from theory and experiments: ● — experimental data, ○ — $r = 0$, △ — $r = 1$, □ — $r = 2$.

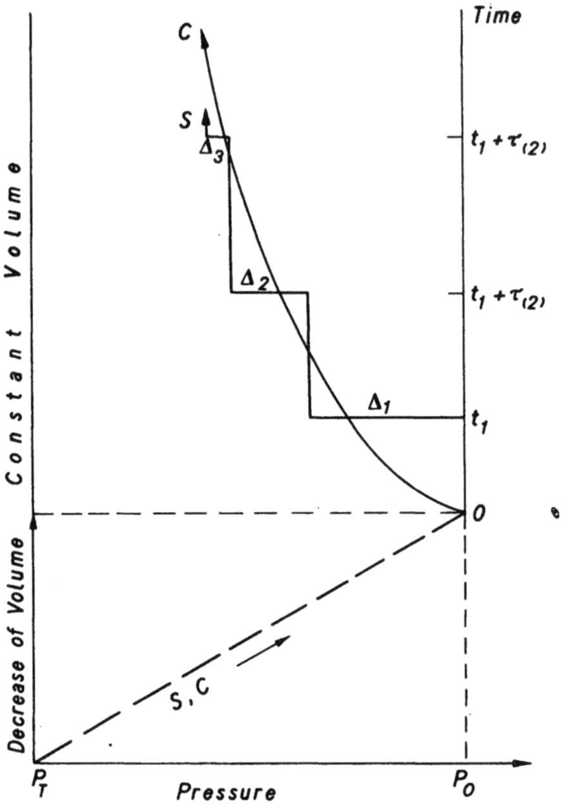

Fig. 4. Theoretical scheme of transformations (the case in which the volume of the high-pressure chamber is constant and pressure exceeds the value of the transformation pressure). s — stepwise transformations, c — continuous transformations; s, c — preliminary stepwise or continuous transformations.

This expression will be used in comparing the theory with the experimental results.

For application of the theoretical model to experiments, the parameters r, s, B and ζ should be calculated from equation (11). In practice the parameters r and s were fixed in advance. For a numerical reason, selection of the parameter r was limited to integers, namely $r = 0, 1$, or 2. For such values of r, the consistency between the theoretical hysteresis curves and experimental data was found in a previous study [281].

For a fixed value of r the probability is independent from parameter B, and it depends on parameter ζ only (see equations (11) and (21)). The latter can be determined from the condition that for the observed transformations their probability is $p_i = 0.5$.

Further procedure is as follows. From consecutive transformations the average value $\bar{\zeta}$ is found, and then this value is introduced into equation (21), from which the probability of the $(i + 1)$-th transformation can be estimated. Fig. 3 shows an example of experimental results obtained for a rubidium chloride and theoretical results calculated with formula (21) in which $v = 0$ was adopted; in other words, we calculated the probability that a transformation will occur in any transformable volume. Times after which one can except the occurrence of a consecutive, j-th stepwise transformation with a probability $p_j = 0.5$

are shown on the axis of ordinates. Each theoretical value was calculated from the observed $(j - 1)$ transformations.

The presented mathematical model could be useful in geophysics for the calculation of earthquake occurrence probability. A calculation technique based on seismis data can be applied to earthquakes. Obviously, the geophysical application of our equations would require the estimation of different specific parameters characterizing an earthquake course. These data are of course less accurate and more difficult to obtain than those from laboratory experiments. It seems, however, that with a sufficiently large number of seismic data the estimation of the necessary parameters could be possible. Thus it is possible to expect that our model could be used for the purpose of earthquake prediction.

Similarly to the presented model of stepwise transformations taking place in the medium whose entire volume decreases, one can theoretically consider transformations taking place in the volume which is constant but subject to pressure $P > P_T$ (Fig. 4). Such a case, which is interesting from the geophysical point of view as a source of earthquakes generated in a region limited by rigid boundaries, has not been observed in the laboratory yet.

SOME ASPECTS OF THE CONNECTION BETWEEN HIGH PRESSURE AND PLANETARY PHYSICS

F. Frölich

As is well known, recently the development of sciences is characterized by a differentiation and integration. A proceeding specialization on the one hand and a gradual growing together on the other are the characteristics of this process of evolution. In this connection, for example the astrophysics and the geosciences are related to one another via the planetary geophysics. Since undoubtedly the Earth has been subject to the most detailed investigations, terrestrial features become the test criteria for every conception of the origin of planets, because it would be absurd to consider the evolution process of the Earth to be a cosmic singularity. From this point of view there are a number of aspects which are of interest for the astrophysics, too.

The geoscientific research requires — since its main field of activity (the Earth's interior) is accessible only by indirect means — a broad "spectrum" of various independent methods of exploration in order to obtain (at least) several data derived along such different lines for each of the different depths:

1. In this connection, the geology as a descriptive geoscience has the task of describing the directly accessible part — the Earth's crust including the eruptively emerged plutonic rocks — in as close a detail as possible with the utilization of mineralogic aspects. Since the geologic exploration is the most obvious kind of activity, till now it provided an abundance of observational results which can be utilized.

2. As a next step it was necessary to explore the structure of the Earth's interior, which already required some amount of instrumentation and for which purpose the seismology was the most suitable means. Since in this case the most important object was to detect discontinuities and inhomogeneities in the matter of the Earth's interior, here — in contrast to the problem of geology — it was just the matter continuum which had to be used as a basis. For a reliable determination of the depths of such anomalies, it was necessary to "filter off" the "noise level" introduced into the seismic wave propagation by the rock varieties.

3. The material-specific aspects, which must necessarily be excluded from the exploration task 2, have to be covered by other means. The seismic method is not suitable for that purpose also because the propagation velocities of the seismic compressional and shear waves (alone) do not allow to draw any conclusions with respect to the rock status, the more so as only two conditional equations are available for the three unknowns: incompressibility (K), density (ϱ), and rigidity (μ). Drawing conclusions will not become

possible until one of the parameters (e.g. K) has been theoretically determined, which is a laborious task because of the great number of varieties of the various rock types.

Here it is more convenient (after the density *vs.* pressure relation has been clarified by means of state equations) to derive, on the basis of theoretical solid state physics, the depth-related regularities of the rock state in the Earth's interior and to examine their effects on silicates, oxides, and sulphides by means of experimental high-pressure physics — say, e.g. by means of laboratory mineral syntheses under high pressure and at high temperatures [284].

Thus the research of details again gains the same importance as in the case of geologic exploration [285, 286]. Supplementary investigations (such as rock-magnetic, geoelectric, geothermal, geochemical, petrologic, gravimetric ones) make their contribution to this and at the same time provide test criteria for other results obtained.

4. Finally, the energy balance, which in this case results in a heat balance, must be taken into consideration as well as the transfer of the heat generated (by radioactive processes) in the Earth's interior, with respect to the three components of heat transfer: conduction, radiation, and advection, the critical question being which peremptory criteria, if any, exist, which presuppose the occurrence of a cold accumulation of the planet, followed by a limited heating-up.

5. The object derived from the above is: a stepwise delimitation of the process of terrestrial origin on the basis of cosmology [287] and with co-consideration of the planetary aspects (accumulation, contraction + thermal processes, redistribution). Since in this case, however, it is the question of an inversion problem, at best a conception of high probability can be achieved, provided that it can be "fitted" to the present overall situation; whereas a physically cogent solution cannot be obtained.

Naturally, all this requires a complex approach, utilizing all reliable data which are available at present, in order to ensure in this way a broad "spectrum" of control parameters. Of the results of geophysical research obtained in this way, of course only a few, essential ones can be described and discussed in this context.

The seismological exploration of the structure of the Earth's interior provided five different depth ranges:
the Earth's crust (down to a depth of 30 km in continental regions,
down to a depth of 5 km in oceanic regions)
the upper mantle (down to a depth of 1000 km),
the lower mantle (down to a depth of 2898 km),
the outer core (down to a depth of 5120 km), and the inner core.

Here the crust differences are of special interest, the more so as they concern not only the thickness, but moreover also the mineral stock as well as the proportion of radioactive constituents.

A hot phase of origin or a complete melting of the Earth's interior necessarily would have led to an (approximately) uniform crust of the Earth.

The altitude frequency curve of the terrestrial topography, which has two peaks, then should have the form of a normal Gauss-distribution.

As it was already mentioned, for a further evaluation it is first necessary to clarify the density *vs.* pressure relationship. The state equations describing the effects of pressure (P) and temperature (T) on the density (ϱ) and the incompressibility

$$K = \varrho/(\partial\varrho/\partial P) = 1/k ,$$

respectively, or its reciprocal value (k) on the basis of the physics of the continuum show that under the conditions of the Earth's interior it is possible to neglect higher-order terms and, for the conditions of the basis of the Earth's mantle, the pressure effects prevail over those of the temperature, since

$$\frac{(\partial K_S/\partial P)_T}{(\partial K_S/\partial T)_P} \approx \frac{25}{1}.$$

Thus for this region the following relations hold:
in the adiabatic case

$$K_S = K_{S_0} + K'_{S_0}P;$$

in the isothermal case

$$K_T = K_{T_0} + K'_{T_0}P.$$

Using the basic assumption that

$$\partial K/\partial P = \text{linear},$$

one obtains

$$\frac{P}{K_{T_0}} = \frac{1}{K'_{T_0}}\left\{\left(\frac{\varrho}{\varrho_0}\right)^{K'_{T_0}} - 1\right\} \qquad \text{(Murnagham)} \qquad (1)$$

In view of the results of complex geophysical research, this latter relation can however be approximately satisfied only in the lower mantle region at depths between 1000 and 2700 km; there by means of the Murnagham-equation the gradient reaches a value of 3.7.

Dropping the assumption of linearity and introducing a new basic assumption

$$\partial K/\partial P = \text{quadratic},$$

one obtains the Birch's state equation

$$\frac{P}{K_{T_0}} = \frac{3}{2}\left\{\left(\frac{\varrho}{\varrho_0}\right)^{7/3} - \left(\frac{\varrho}{\varrho_0}\right)^{5/3}\right\}\left\{1 - \underbrace{\frac{3}{4}(4 - K'_{T_0})}_{\xi}\left[\left(\frac{\varrho}{\varrho_0}\right)^{2/3} - 1\right]\right\}. \qquad (2)$$

For $\xi = 3 - 3K'_{T_0}/4 \to 0$ one obtains $K'_T \to 4$, which is in approximate agreement with the results of observations made at the Earth's surface $(K'_{T_0} \lessgtr \frac{5}{4})$.

However, for the basis of the Earth's mantle the results are for

$$\begin{aligned} \xi &= +1/2: & \partial K/\partial P &= 1.40, \\ \xi &= 0: & \partial K/\partial P &= 2.85, \\ \xi &= -1/2: & \partial K/\partial P &= 3.34. \end{aligned}$$

This uncertainty was the cause of further trials:

By way of example, if structural variations are taken into account and hence the considerations are based on the ionic-lattice potential and its terms for attraction ($\varkappa$) and repulsion (ν), one obtains

$$\frac{P}{K'_{T_0}} = \frac{3}{\nu - \varkappa}\left\{\left(\frac{\varrho}{\varrho_0}\right)^{(\nu+3)/3} - \left(\frac{\varrho}{\varrho_0}\right)^{(\varkappa+3)/3}\right\} \qquad \text{(Anderson)}. \qquad (3)$$

Thus for the conditions near the Earth's surface,

$$\varrho/\varrho_0 \gtrapprox 1; \quad \nu = 7,$$

this results in a better agreement with the results of observation, i.e.

$$K'_{T_0} = 4.203;$$

on the other hand, at the basis of the Earth's mantle one has for non-ferreous oxides

$$K'_T = 3.33,$$

whereas for ferreous oxides

$$K'_T = 3.67,$$

because here the repulsion should dominate as a consequence of high pressure.

Another way involves increasing the number of parameters for fitting the relation to the results of evaluation of shock-wave investigations. The expression extended in this way,

$$\frac{P}{K_{T_0}} = \frac{K'_{T0}}{K'^2_{T_\infty}} \left\{ \left(\frac{\varrho}{\varrho_0}\right)^{K'_{T\infty}} - 1 \right\} - \frac{K'_{T0} - K'_{T\infty}}{K'_{T_\infty}} \ln\left(\frac{\varrho}{\varrho_0}\right) \qquad \text{(Keane)} \qquad (4)$$

is based on the basic assumption that

$\partial K/\partial P$ is monotonically decreasing;

passing to the linearity assumption,

$$K'_{T_\infty} = K'_{T_0}$$

yields the Murnagham equation, where however K'_{T_∞} generally is less than K'_{T_0}. The difference between (4) and (1) is 15% at the basis of the Earth's mantle; between (4) and (3), however, it is as low as 1.5%, reaching this value only at extreme pressure values.

Now the extensive results of the international UPPER-MANTLE-PROJECT [288, 289] show, among other facts, that olivine,

$$\{x\, \underset{\text{forsterite}}{Mg_2SiO_4}(1-x)\, \underset{\text{fayalite}}{Fe_2SiO_4}\}$$

is the dominant constituent of the upper mantle. As is shown by the simulation of the *PT* conditions in laboratory experiments, this olivine undergoes a phase change — caused by high pressure — from the original neosilicate structure to the spinel structure, which effects a compaction by more than 10%. Naturally such a jump-like transition to a higher density phase results in a liberation of heat, which in turn necessarily causes an increase in the electrical conductivity in this semiconductor. This was confirmed by experiment, too; in fayalite the conductivity increases by two orders of magnitude [155]. We shall return to this fact when discussing the thermal problems.

Further investigations under static high pressure show that quartz (SiO_2) is converted into a different phase — the so-called stishovite — whereby the number of the nearest lattice neighbours, which usually amounts to 4, is increased to 6 [290]. Since however the normal, so-called 4-coordination together is the construction-unit of silicate structures (i.e. the neosilicates, chain-, ring-, band-, sheet-, and frame-work-type silicates [291]), these possibilities then are not applicable, so that for the material stock at greater depths in effect only metal oxides will be left in addition to stishovite and iron sulphides [292, 293, 294].

The same result is obtained also by means of other investigations than those stated here, e.g. by studying geochemical and other relationships being independent of the above.

By the way, these two conversions caused by pressure are of special importance in the course of terrestrial evolution; later on we shall return to this problem.

For the core of the Earth, the seismological exploration shows that the shear waves in the outer core region disappear, which is equivalent with postulating a fluid state in this region, because only in this case the rigidity (μ), to which the velocity of propagation of the shear waves is proportional, becomes zero. In the inner part of the core, however, shear waves are again found evident; consequently this part is again solid, since at the phase boundaries usually a change of the wave mode (compressional to shear waves) occurs.

The other available test criteria, being established on the basis of the complex approach of geoscientific research as outlined at the beginning of this paper, are stated below:

— the density ratio, $\varrho_{core}/\varrho_{mantle} = 9.7/5.5 = 1.76$,
— the discontinuous increase of electric conductivity:

$$\sigma_{semiconductor} \rightarrow \sigma_{metallic} ,$$

— the mean atomic number $\bar{Z} \begin{smallmatrix} < 26 \\ > 20 \end{smallmatrix}$,
— the detailed results of meteorite research,
— the "seismological sharp" boundary between the core and the mantle, with a possibility of lateral (topographic) deviations along its course,
— the content of radioactive matter,
— the possibility of a thermally caused turbulence (i.e. the elimination of the Higgins-Kennedy paradox [295, 296, 297] as a condition of origin of the main magnetic field,
— the *PT*-laws, which can be derived from solid-state physics, taking into account the effects of the crystal field and the variations in the bonding state [298, 299].

As a supplement to this, meanwhile the experimental high-pressure physics has shown that in the (Mbar) pressure range to be expected at the core/mantle boundary oxides of transition metals (α — Fe_2O_3, TiO_2, Cr_2O_3, NiO) are "metallized" during a change of structure from the oxide structure (NaCl) to the body-centred cubic structure of CsCl [300].

As regards the thermal side of the exploration tasks, the corresponding detailed investigations [301—307] show that the heat liberated by the decay of the radioactive constituents alone (33×10^{36} erg) is sufficient for effecting the observable, thermally caused phenomena.

This heat liberated in the Earth's interior is transferred towards the Earth's surface mainly by heat-conduction and heat-radiation processes. The negative temperature dependence of the contribution of heat-conduction, which is due to the thermal expansion of the solid, is proceedingly decreased and finally overcompensated by the rising pressure as the depth increases. The minimum of the heat conduction effects lies at a depth of about 180 km, i.e. about there where the pressure-induced phase change of the olivine begins and fayalite is converted into the spinel phase (as was already mentioned). On the other hand, from the Earth's surface downwards the contribution of heat-radiation increases very rapidly as a consequence of its strong positive temperature dependence, representing the dominant portion of heat transfer at the above mentioned depth of 180 km. It can be limited only by an increasing of electrical conductivity.

Now, since the pressure-induced conversion of the olivine into the more compact spinel phase naturally leads to a liberation of heat and an evidently increasing of the electrical conductivity (as stated above) by two orders of magnitude, a thermal shielding of the heat radiation — prevailing especially in this region — must necessarily be released there, being additionally supported by the heat conduction, which is at a minimum in this region [308].

Since on the other hand the phase transition does not occur in the whole mixing range of the olivine but, as it is shown by high-pressure experiments, only in the range from pure fayalite to 70 mole % of forsterite, only a regionally limited shielding of the heat transfer from the Earth's interior can be expected:

forsterite	fayalite
x	$(1 - x)$

x: 0 ... 0.7 spinel phase (shielding)
0.7 ... 0.8 transition (partially penetrable)
0.8 ... 1.0 (penetrable for heat radiation)

We compare this finding with the situation of the terrestrial surface regions:

	%	10^6 km²	%
Region of continental shields	20.6	105	29.2
young zones of folding	8.2	42	
volcanic islands	0.4	2	
shelf and continental slope zones	18.2	93	70.8
deep sea region	52.6	268	
	100.0%	510	

Here a conspicuous agreement is observed, which satisfies the "fitting" requirement and the test criteria in an instructive way. Therefore, as a consequence of the continual ascending heat transfer, a melting can occur in the terrestrial surface region above the region with the highest forsterite content, followed by a gravitational and radioactive differentiation. However, due to the relative high viscosity of the little by little molten phase, in which this process must occur, such a separation of the lighter constituents from the heavier ones is a highly delayed process, which requires a considerable time.

On the basis of these and other results of the interdisciplinary geoscientific research — which cannot be described here in greater detail, because this is beyond the scope of this paper — now also the process of planetary origin can be subjected to a closer investigation:

Hypotheses — in any case being legitimate means of scientific research — become practically indispensable when (whatever the cause may be) there is no possibility of clarifying a problem by experiment. The value of such a hypothesis depends on

- whether all of the observational findings available for the respective problem can be derived from this hypothesis and
- whether further implications of the hypothesis lead to new knowledge.

Both of these criteria are met by the conception of O. Yu. Schmidt [309—312].

If the whole set of reliably detected phenomena of our planetary system as well as the results obtained in the terrestrial case are compared with the hypotheses existing at present, it turns out that this conception alone ensures that the above criteria are satisfied — including the second criterion, the more so as at the time of its publication the thermal history of the Earth, for one example, still was an entirely open problem.

The basic postulates of this conception of a cold accumulation comprise only the following two obvious considerations:

— Within an extensive particle swarm near the sun, particles move — due to gravitational forces — along elliptic orbits around the sun, as a consequence of a non-zero angular momentum.
— This results in collisions and particle aggregations. (As regards the latter, the coagulating effect of moisture (water, ammonia) as a contributing factor was subsequently taken into account by Urey [313] from a physico-chemical point of view.)

In any case the complanarity of the planet orbits and the equal directions of revolution in our planetary system indicate the pre-existence of such an angular momentum, since — according to the principle of conservation of angular momentum — the latter could not arise during the process of planetary evolution. The almost circular planetary orbits result as an effect of averaging from the eccentric particle orbits and therefore depend on the planet size; this averaging reaches its optimum for the giant planets [314, 315].

On the basis of this conception and the results obtained from the interdisciplinary terrestrial research, to which contributions were made by all branches of geosciences, it is possible to derive a course of geo-formation in several successive stages:

— With increasing size of the planetary solid body there was a variation in the pressure and temperature conditions in the strate far from the surface, the rock matter had to change with increasing inclusion in the interior of the planet, because the equilibrium situation determining the rock type of the original surface layer was no longer existent. This resulted in pressure-induced phase transitions and, with increasing depth, in a change of the material composition (silicates → oxides).
— As the pressure in the interior increases (during the phase of accumulation of the planetary solid body), the "metallization" in the core of the Earth is released when the Mbar — range is reached, followed by the conversion into the fluid state — which first occurs within the "metallized" zone because of the diminished cohesion energy of the metallic state — under the full influence of the proceeding heating-up by the decay of the radioactive constituents. The term "fluid" shall indicate that also some small portions of finest-grained solid particles may be co-existent in the melt with a depth-dependent different distribution, which — at least for the time being — cannot be excluded (and which correspond to the elimination of the above mentioned Higgins-Kennedy-paradox).
— A further pressure increase in the course of proceeding accumulation results in a re-"metallization" in the interior of the core region.

With this attempt at a reconstruction of the course of evolution it is possible to achieve full agreement with all of the results of exploration which are available at present, including the material-specific aspects and the various couplings between the core and the mantle.

Naturally, the pressure-induced "metallization" is influenced by differences in the composition of the oxides at the basis of the mantle, so that lateral (topographic) deviations in the course of the core/mantle boundary may develop, for which there are a number of indications; however, this boundary should always be "seismological sharp" — at whatever the exact depth may be — in agreement with the (above mentioned) results of seismological exploration.

Finally, the pre-conditions of the generation of the geomagnetic main field are ensured, too.

For all that, of course this attempt at a solution is not physically cogent — as it was emphasized already at the beginning of this paper — but at best probable, since the "fitting" requirement is satisfied, at least on the basis of the hitherto available data, which does not, however, exclude any modifications due to future, new experimental results.

Using this (self-contained and consistent) conception — further details will be published in the scientific series of the Central Earth Physics Institute Potsdam — as a working hypothesis, it is possible to carry on the investigations about the origin of the Earth/Moon system, where it should be taken into account that the heat balance of the two planetary solids is increased by the conversion of rotational energy of the Earth as a consequence of the lunar-terrestrial tidal friction.

The explorations from this point of view are carried on also for the group of earth-similar planets.

References

[1] BAYDYUK, B. V.: Mechanical properties of rocks under high pressures and temperatures (in Russian). Gostoptekhizdat, 1963.

[2] BAYUK, E. I.; TEDEEV, A. V.: The temperature effect on longitudinal velocities of rocks under high pressures (in Russian). In: Physical properties of rocks at high thermodynamic parameters. Naukova dumka, Kiev, 7—10.

[3] BRADLY, K.: Application of high pressure technique to solids studies (Russian translation). Mir, Moscow 1972.

[4] DAVIDSON, G. E.; KENDALL, D. P.: Design and construction of pressure vessels (Russian translation). In: Mechanical properties of minerals under high pressure. Mir, Moscow 1973, 139—147.

[5] GONIKBERG, M. T.: High and super-high pressures in chemistry (in Russian). Nauka, Moscow 1968.

[6] LEBEDEV, T. S.; SHAPOVAL, V. I.; KORCHIN, V. A.: Investigations of the high pressure and temperature effect on the velocity change in igneous and metamorphic rocks (in Russian). Geofiz. sb. AN Ukr. SSR (1969) 31, 75—93.

[7] LEBEDEV, T. S.; SHAPOVAL, V. I.; KORCHIN, V. A.: Investigation of high pressure and temperature effect on the elastic wave velocity in some rocks of the Ukrainian Shield (in Russian). Geofiz. sb. AN SSR (1969) 32, 5—17.

[8] LEBEDEV, T. S.; SHAPOVAL, V. I.; KORCHIN, V. A.: Studying of some elastic parameters of rocks under high pressures and temperatures (in Russian). In: Physical properties of rocks at high thermodynamic parameters. Naukova dumka, Kiev 1971, 33—36.

[9] LEBEDEV, T. S.; et al.: Elastic properties of rocks under high pressures (in Russian). Naukova Dumka, Kiev 1972.

[10] LEBEDEV, T. S.; SHAPOVAL, V. I.; KORCHIN, V. A.: New data of longitudinal velocities in rocks under high thermodynamic parameters (in Russian). Geofiz. sb. AN Ukr. SSR (1972) 49, 9—27.

[11] LEBEDEV, T. S.; SHAPOVAL, V. I.; KORCHIN, V. A.; LAKHTANOV, V. T.: Methods and apparatus for simultaneous studies of transverse and longitudinal velocities at high pressures (in Russian). Geofiz. sb. AN Ukr. SSR (1974) 58, 41—49.

[12] LEBEDEV, T. S.: Physical properties of the lithosphere rocks under conditions of the deep interior (in Russian). In: The problem of physics of the Earth in the Ukraine. Naukova Dumka, Kiev 1975, 98—117.

[13] LEBEDEV, T. S.; KORCHIN, V. A.; VASILYAKA, V. T.: Hydrostatic investigations of elastic properties of mineral matter under high thermodynamic parameters (in Russian). Geofiz. sb. AN Ukr. SSR (1976) 69, 26—34.

[14] LEBEDEV, T. S.; SHAPOVAL, V. I.; KORCHIN, V. A.: Elastic velocities in rocks at high values of thermodynamic parameters. Zeitschr. geol. Wiss. **1** (1973) 4, 431—450.

[15] LEBEDEV, T. S.; SHAPOVAL, V. I.; KORCHIN, V. A.: Programmed modelling of deep thermodynamic conditions in experimental studies of elastic properties of mineral matter. Gerlands Beitr. z. Geophysik **83** (1974) 5, S. 340—346.

[16] LEBEDEV, T. S.: Investigation of some physical properties of mineral matter under thermodynamic conditions of the Earth's interior. Zeitschr. geol. Wiss. **3** (1975) 6, 717—739.

[17] Modern Super-High Pressure Techniques (Russian translation). Mir, Moscow 1964.

[18] OGIBALOV, P. M.; KIYKO, I. A.: Essays on mechanis of high parameters (in Russian). Izv. MGU, Moscow 1966.

[19] POPOVA, S. V.; BENDELIANY, N. A.: High pressures (in Russian). Nauka, Moscow 1974.

[20] RUMANS, K.: The structural changes in oxides and other chalcogenides under normal and high pressures (Russian translation). Mir, Moscow 1969.

[21] RYAPOSOV, A. P.; KIRKINSKY, V. A.: High temperature heating in hydrostatic pressure chambers (in Russian). In: Experimental studies in mineralogy. Novosibirsk 1969, 164—167.

[22] SHITYAEV, A. YA.: Phase transforms and properties of alloys under high pressures (in Russian). Nauka, Moscow 1973.

[23] SVENSON, K.: High pressure physics (Russian translation). IL., Moscow 1963.

[24] TSYKLIS, D. S.: High pressure techniques of physical-chemical investigations (in Russian). Goskhimizdat, Moscow 1951.

[25] MCSKIMIN, H. J.: Elastic moduli of single crystal Germanium as a function of hydrostatic pressure. J. Acoust. Soc. Am. **30** (1958) 314—318.

[26] MCSKIMIN, H. J.: Measurement of elastic constants at low temperatures by means of ultrasonic waves data for Silicon and Germanium single crystals, and for fused Silica. J. Appl. Phys. **24** (1953) 988—997.

[27] BOLEF, D. I.; MENES, M.: Nuclear magnetic resonance acoustic absorption in KJ and KBr. Phys. Rev. **114** (1959) 1441—1451.

[28] BOLEF, D. I.; DE KLERK, J.: Some continuous wave-techniques for the measurement of velocity and attenuation of ultrasonic waves between 1 and 1000 MHz. IEEE Transactions on Ultrasonic Engineering, **UE 10** (1963) 19—26.

[29] LEVYKIN, A. I.: Absorption and velocity of compressional and shear waves in rocks under hydrostatic pressure to 4000 kg/cm^2. Izv. AN SSSR, Ser. fiz. Zemli (1965) 2, 21—27.

[30] LEBEDEV, T. S.; LAKHTANOV, U. T.; PRODAJVODA, G. T.: Measurement of the propagation velocity and absorption of ultrasonic waves in minerals and rocks by a pulseecho method. Geofiz. sb. AN Ukr. SSR (1971) 43, 63—67.

[31] WAGNER, F. C.: Zur Verwendung eines Fehlerortungsgerätes für Ultraschalluntersuchungen. Bergakademie **10** (1967) 623.

[32] CEDRONE, N. P.; CURRAN, D. R.: Electronic pulse methode for measuring the velocity of sound in liquids and solids. J. Acoust. Soc. Am. **26** (1954) 963—966.

[33] AKULOVA, R. S.; BAYUK, E. I.: The velocity of elastic waves in halolites at pressures up to 24 kbars (in Russian). In: Physical properties of rocks under high pressures and temperatures. Mietsniereba, Tbilissi 1974, 63—65.

[34] BAYUK, E. I.: The effects of high pressure on the velocity of longitudinal waves in salt and other sedimentary rocks of the Prinyat depression (in Russian). Izv. AN SSSR, Ser. fiz. Zemli (1969) 3, 24—29.

[35] VOLAROVICH, M. P.; BAYUK, E. I.; TEDEEV, P. V.: Study of the elastic properties of rock salt at high uniform pressures up to 4 kbars (in Russian). Dokl. AN SSSR (1972) 1, 56—59.

[36] VORONOV, F. F.; GONCHAROVA, B. A.; GRIGORYEV, S. B.: The effects of pressure up to 20 kbars on the elastic characteristics of chlorides (in Russian). Fiz. tverdogo tela **13** (1971) 5, 1353—1356.

[37] BAYUK, E. I.; EPINATEVA, A. M.; KARUS, E. B.; SHKERINA, L. P.: The velocity of longitudinal waves under the conditions of natural rock deposition and on samples at high pressures (in Russian). Izv. AN SSSR, Ser. fiz. Zemli, (1974) 2, 94—99.

[38] VOLAROVICH, M. P.; LEVYKIN, A. I.; GALDIN, N. E.: Study of the velocities of longitudinal waves in rock samples at pressures up to 20000 kg/cm (in Russian). Dokl. AN SSSR **157** (1964) 6, 1349—1351.

[39] BAYUK, E. I.; AKULOVA, F. S.: Study of polymorphism in potassium halogenides at high pressures by ultrasonic method (in Russian). Dokl. AN SSSR **227** (1976) 6, 1342—1344.

[40] AFANASYEV, G. D.; BAYUK, E. I.; BELIKOV, B. P.; VOLAROVICH, M. P.; YEFIMOVA, G. A.; LEVYKIN, A. I.: Investigation of elastic velocities in some rock-forming minerals under pressures to 20 kbar (in Russian). Dokl. AN SSSR **201** (1971) 3, 586—589.

[41] ANDERSON, O. L.: Determination and some applications of isotropic elastic constants of polycrystalline systems based on single crystal data. In: Lattice dynamics (Russian transl.) Mir, Moscow 1968, 62—122.

[42] BELIKOV, B. P.; ALEXANDROV, K. S.; RYZHOVA, T. V.: Elastic properties of rock-forming minerals and rocks (in Russian). Nauka, Moscow 1970.

[43] FRIEDEL, L.: Dislocations (in Russian). Mir, Moscow 1967.

[44] INDENBOM, V. L.; ORLOV, A. N.: Physical theory of plasticity and strength (in Russian). Uspechi fiz. Nauk **26** (1962) 3, 557—591.

[45] KELLY, A.; GROWS, G.: Crystallography and defects in crystals (in Russian). Mir, Moscow 1974.

[46] KORCHIN, V. A.; RUBAN, V. A.; DANILOV, V. V.: Reduction of the width of FMR under effect of quasihydrostatic pressure in single crystals of ferrite (in Russian). Fiz. tverdogo tela **13** (1971) 7, 2124; 2125.

[47] KUZMENKO, V. A.: New patterns of solids deformations (in Russian). Naukova dumka, Kiev 1973.

[48] LEBEDEV, T. S.; OROVETSKY, YU. P.; KORCHIN, V. A.: High hydrostatic pressure studies of the changes in quartz dislocational structure and their interrelation with physical properties of quartz-bearing rocks (in Russian). Geofiz. sb. AN Ukr. SSR (1976) 70, 71—85.

[49] LEBEDEV, T. S.; POZNANSKAYA, N. F.; SAVENKO, B. Y.; SHAPOVAL, V. I.; KORCHIN, V. A.: Investigations of the effect of high hydrostatic pressures and temperatures on the remanent magnetization of magnetite and magnetite-bearing rocks (in Russian). Geofiz. sb. AN Ukr. SSR (1973) 56, 41—49.

[50] MCCLEANTOCK, F.; ARGON, A.: Deformation and destruction of materials (in Russian). Mir, Moscow 1970.

[51] PINES, B. Y.: Diffusion and mechanical properties of solids (in Russian). Uspechi fiz. Nauk 26 (1962) 3, 519—556.

[52] ... Physics of deformational strengthening of single crystals (in Russian). Naukova Dumka, Kiev 1972.

[53] PSHENICHKOV, Y. P.: Defection of fine structure of crystals (in Russian). Metallurgiya, Moscow 1974.

[54] RADCLIFFE, S. V.: Effect of hydrostatic pressure on structural defects and properties. In: Mechanical properties of materials under high pressure, I (in Russian). Mir, Moscow 1973, 254—295.

[55] RYZHOVA, T. V.: Elastic properties of plagioclases (in Russian). Izv. AN SSSR, Ser. geofiz. (1964) 7, 1049—1051.

[56] VAN BÜREN: Defects in crystals (in Russian). IL, Moscow 1960.

[57] VOLAROVICH, M. P.; BAYUK, E. I.; LEVYKIN, A. I.; TOMASHEVSKAYA: Physical-mechanical properties of rocks and minerals at high pressures and temperatures (in Russian). Nauka, Moscow 1974.

[58] VOLAROVICH, M. P.; BAYUK, E. I.; YEFIMOVA, G. A.: Elastic properties of minerals at high pressures (in Russian). Nauka, Moscow 1975.

[59] YUSHKIN, YU. P.: Mechanical properties of minerals (in Russian). Nauka, Leningrad 1971.

[60] ZHARKOV, V. P.; KALININ, V. A.: Equation of state for a solid at high pressures and temperatures (in Russian). Nauka, Moscow 1968.

[61] SOBOLEV, V. S.; SOBOLEV, N. V.: Xenolithes in Kimberlites from north Yakutia and problems of the Earth mantle structure (in Russian). Dokl. AN SSSR **158** (1964).

[62] GENSHAFT, YU. S.: Experimental model of mineral composition of Earth core and mantle (in Russian). In: Physical properties, composition and structure of upper mantle. Nauka, Moscow 1974.

[63] GREEN, D. H.; RINGWOOD, A. E.; et al.: Petrology of the upper mantle (Russian transl.). Mir, Moscow 1968.

[64] VOLAROVICH, M. P.; BAYUK, E. I.; YEFIMOVA, G. A.: Elastic properties of minerals at high pressures (in Russian). Nauka, Moscow 1975.

[65] CHRISTENSEN, N.; RAMANANANTOANDRO, R.: Elastic moduli and anisotropy of dunite to 10 kbar. J. geophys. Res. **76** (1971) 4003—4009.

[66] LEVYKIN, A. I.: Elastic properties of minerals and monomineral rocks at pressures up to 55 kbar (in Russian). In: Physical properties of rocks at high pressures and temperatures. Mitsniereba, Tbilissi 1974, 90—92.

[67] CHUNG, D. H.: Effect of iron/magnesium ratio on p- and s-wave velocities on olivine. J. geophys. Res. **75** (1970) 7353—7361.

[68] HALLECK, P. M.: The compression and compressibility of grossular garnet: a comparison of X-ray and ultrasonic methods. Ph. D. thesis, Univ. of Chicago, Chicago 1973.

[69] VOLAROVICH, M. P.; LEVYKIN, A. I.: Influence of pressure up to 40 kbar on the elastic properties of rock-forming minerals (in Russian). Geofiz. sb. AN Ukr. SSR (1973) 54, 35—38.

[70] SOGA, N.: Elastic constants of garnet under pressure and temperature. J. geophys. Res. **72** (1967) 4227—4234.

[71] WANG, H.; SIMMONS, G.: Elasticity of some mantle crystal structures, 3, spessartite-almandine garnet. J. geophys. Res. **79** (1974) 2607—2613.

[72] BIRCH, F.; BANCROFT, D.: New measurements of the rigidity of rocks at high pressure. J. Geol. **48** (1940) 752—768.

[73] HUGHES, D. S.; JONES, H. J.: Variations of elastic moduli of igneous rocks with pressure and temperature. Bull. geol. Soc. Amer. **61** (1950) 843—856.

[74] HUGHES, D. S.; CROSS, J. H.: Elastic wave velocities in rocks at high pressures and temperatures. Geophysics **16** (1951) 577—593.

[75] HUGHES, D. S.; MAURETTE, C.: Variations of elastic wave velocities in granites with pressure and temperature. Geophysics **21** (1956) 277—284.

[76] KING, M. S.; FATT, J.: Ultrasonic shear wave velocities in rocks subjected to simulated overburden pressure. Geophysics **27** (1962) 590—598.

[77] WILLIE, M. R. J.; GARDNER, G. H. F.; GREGORY, A. R.: Studies of elastic wave attenuation in porous media. Geophysics **27** (1962) 569—589.

[78] BIRCH, F.: The velocity of compressional waves in rocks to 10 kilobars. J. geoph. Res. **65** (1960) 1083—1102.

[79] BIRCH, F.: The velocity of compressional waves in rocks to 10 kilobars. J. geoph. Res. **66** (1961) 2199—2224.

[80] SIMMONS, G.: Velocity of shear waves in rocks to 10 kilobars. J. geoph. Res. **69** (1964) 1123—1130.

[81] HADLEY, K.: Comparison of calculated and observed crack densities and seismic velocities in Westerly granite. J. geoph. Res. **81** (1976) 3484—3494.

[82] AFANASYEV, G. D.: Structure and composition of the Earth's crust as related to problems of geologic petrology (in Russian). Izv. ANSSSR, Ser. geol. **2** (1970) 3—26.

[83] BAYUK, E. I.: Elastic properties of minerals under high pressures and temperatures (in Russian). Geofiz. sb. AN Ukr. SSR (1973) 51, 27—32.

[84] BAYUK, E. I.; TEDEYEV, R. V.: Longitudinal wave velocity in rock samples under a simultaneous high pressure and temperature effect (in Russian). Izv. AN SSSR, Fiz. Zemli (1974) 8, 63—70.

[85] BELOUSSOV, V. V.: The Earth's crust and upper mantle of the continents (in Russian). Nauka, Moscow 1966.

[86] BIRCH, F.: Elasticity of igneous rocks at high temperatures and pressures. Bull. Geol. Soc. Amer. **54** (1943) 263—286.

[87] BIRCH, F.: Interpretation of the seismic structure of the crust in the light of experimental studies of wave velocities in rocks. Contributions in Geophysics in Honor of Beno Gutenberg, Pergamon Press New York, **1** (1958) 158—170.

[88] BIRCH, F.; BANCROFT, D.: The elasticity of glass at high temperatures and the vitreous basaltic substratum. Amer. J. Sci. **240** (1942) 457—490.

[89] DMITRIEV, A. P.: Physical properties of rocks under high temperatures (in Russian). Nedra, Moscow 1969.

[90] FIELITZ, K.: Elastic wave velocities in different rocks of high pressure and temperatures up to 750 °C. Zeitschr. f. Geophysik **37** (1971) 6, 943—956.

[91] Physical acoustics (Mason, Editor) — Russian translation, I. Mir, Moscow 1966.

[92] Physical properties of rocks under high thermodynamic parameters (in Russian). Proc. of the III. All-Union Meeting. Naukova Dumka, Kiev 1971.

[93] Physical properties of rocks under high pressures and temperatures (in Russian). Proc. of the IV. All-Union-Meeting. Metsniereba, Tbilissi 1974.

[94] GZOVSKY, M. V.: Fundamentals of tectonophysics (in Russian). Nauka, Moscow 1975.

[95] Handbook of Physical Constants (Editor — Sydney P. Clark, Jr.). Yale University, New Haven, Connection. Published by the society, 1966.

[96] HUGHES, D. S.; MAURETTE, C.: Variation of elastic wave velocities in basic igneous rocks with pressure and temperature. Geophysics **22** (1957) 23—31.

[97] HUGHES, D. S.; MAURETTE, C.: Determination des vitesses d'onde élastique dans diverses roches en fonction de la pression et de la température. Rev. Inst. Franc. Petrole et Ann. combustible, liquide **12** (1957) 730—752.

[98] KUZ'MENKO, V. A.: New schemes of solids deformation (in Russian). Naukova Dumka, Kiev 1973.

[99] LANDAU, L. D., LIFSHITS, E. M.: Theory of elasticity (in Russian). Nauka, Moscow 1965.

[100] LEBEDEV, T. S.; BURTNY, P. A.; KORCHIN, V. A.: Interrelation between pore space and elastic velocity in rocks under high pressure (in Russian). Geofiz. sb. AN Ukr. SSR (1973) 54, 7—17.

[101] LEBEDEV, T. S.; BURTNY, P. A.; KORCHIN, V. A.: Behaviour of the pore space in granits and its relationship with elastic wave velocities at high pressure and temperatures. Gerlands Beitr. z. Geophysik **83** (1974) 170—181.

[102] LEBEDEV, T. S.; BURTNY, P. A.; KORCHIN, V. A.: Certain problems of the porous space behaviour and its relation to elastic wave velocity in rocks under high pressure (in Russian). In: Geophysical studies of the Earth's crust and mantle under the Ukraine. Naukova Dumka, Kiev 1975, 46—51.

[103] LEBEDEV, T. S.; SHAPOVAL, V. I.; KORCHIN, V. A.: Effect of high temperature and pressure on changes in ultrasonic velocities in some igneous and metamorphic rocks of the Ukrainian Shield. Gerlands Beitr. z. Geophys. **79** (1970) 108—116.

[104] LEBEDEV, T. S., SHAPOVAL, V. I.; KORCHIN, V. A.: A study of some elastic parameters of rocks at high pressures and temperature. Report to XV. General Assembly of the IUGA (Moscow, 30. July to 14. August 1971). Naukova Dumka, Kiev 1971, 38—41.

[105] LEBEDEV, T. S.; SHAPOVAL, V. I.; KORCHIN, V. A.: Untersuchungen der Geschwindigkeit von elastischen Wellen in Gesteinen unter den thermodynamischen Bedingungen der Erdkruste. Veröffentl. Zentralinst. Physik d. Erde **22** (1974) 17—34.

[106] LEBEDEV, T. S.; OROVETSKY, YU. P.; SHAPOVAL, V. I.; KORCHIN, V. A.: Polymorphic transitions in carbonaceous rocks and ultrasonic velocities in these and certain igneous and metamorphic formations under hydrostatic pressure to 15 kbar (in Russian). Geofiz. sb. AN Ukr. SSR (1974) 62, 57—63.

[107] LEBEDEV, T. S.; OROVETSKY, YU. P.; KORCHIN, V. A.: The change of the morion crystal structure under hydrostatic pressure to 15 kbar (in Russian). In: Theses of the reports to the V. International Conference on the high pressure physics and technique. Nauka, Moscow (1975) 214.

[108] LEBEDEV, T. S.; OROVETSKY, YU. P.; SHAPOVAL, V. I.; KORCHIN, V. A.: Ultrasonic velocities and polymorphic transitions in carbonaceous rocks under hydrostatic pressure to 15 kbar (in Russian). In: Physical properties of rocks under high pressures and temperatures. Metsniereba, Tbilissi 1974, 83—86.

[109] LEBEDEV, T. S.; OROVETSKY, YU. P.; KORCHIN, V. A.: Dynamics of polymorphous transitions and ultrasonic velocity variation in carbonaceous rocks and mineral with $CaCO_3$ in various thermodynamic conditions (in Russian). In: Investigations of physical properties of mineral matter of the Earth's under high thermodynamic parameters. Naukova Dumka, Kiev 1977 (in press).

[110] LEBEDEV, T. S.; VOLLSTAEDT, H.; KORCHIN, V. A.; SEIPOLD, U.; SHAPOVAL, V. I.; FREUND, D.: Comparative characteristics of methods for studying certain physical parameters of rocks under varying *PT*-regimes in experiments conducted in GDR and USSR (in Russian). In: Physical properties of rocks under high pressures and temperatures. Metsniereba, Tbilissi 1974, 42—45.

[111] LEBEDEV, T. S.; SHAPOVAL, V. I.; KORCHIN, V. A.: Experimental studies of the effect of high temperatures and pressures on the velocity variations in igneous rocks of the Ukrainian Shield (in Russian).

In: Pressure and mechanical stresses in development of the composition, structure and topography of the lithosphere. Nedra, Leningrad 1969, 216—218.

[112] LEBEDEV, T. S.; SHAPOVAL, V. I.; KORCHIN, V. A.; BURTNY, P. A.: Experimental data on certain physical properties of the crystalline rocks from the central part of the Ukrainian Shield (in Russian). Geofiz. sb. AN Ukr. SSR (1970) 34, 19—28.

[113] LEBEDEV, T. S.; SHAPOVAL, V. I.; KORCHIN, V. A.: Physical parameters of certain granitoids in experimental conditions corresponding to the upper crust (in Russian). In: Physical properties, composition and structure of the upper mantle. Moscow 1971, 32—33.

[114] LEBEDEV, T. S.; SHAPOVAL, V. I.; KORCHIN, V. A.: Deformation and density characteristics of magmatic and metamorphic rocks under high pressure (in Russian). In: Physical properties of rocks under high thermodynamic parameters. Naukova Dumka, Kiev 1971, 190—194.

[115] LEBEDEV, T. S.; SHAPOVAL, V. I.; KORCHIN, V. A.: Investigation of certain elastic parameters of rocks under high pressures and temperatures. (ibidem) 33—36.

[116] LEBEDEV, T. S.; SHAPOVAL, V. I.; KORCHIN, V. A.: Elastic properties of rocks of the Ukrainian Shield under high temperatures (experimental data). (ibidem) 128—131.

[117] LEBEDEV, T. S.; SHAPOVAL, V. I.; KORCHIN, V. A.: A device for the studies of physical properties of rocks under high pressure to 15 kbar and high temperature (in Russian). Geofiz. sb. AN Ukr. SSR (1974) 59, 30—35.

[118] LEBEDEV, T. S.; SHAPOVAL, V. I.; KORCHIN, V. A.: Investigation of physical properties of rocks and minerals under pressure to 15 kbar and high temperatures (in Russian). In: Physical properties of rocks under high pressures and temperatures. Metsniereba, Tbilissi 1974, 87—89.

[119] LEBEDEV, T. S.; SHAPOVAL, V. I.; KORCHIN, V. A.: Modelling of the regional thermodynamic conditions of the deepseated crust in studying physical properties of mineral matter. (ibidem) 40—41.

[120] LEBEDEV, T. S.; SHAPOVAL, V. I.; KORCHIN, V. A.: Programmed modelling of deep thermodynamic conditions in experimental studies of elastic properties of mineral matter (in Russian). In: Geophysical studies of the Earth's crust and mantle under the Ukraine. Naukova Dumka, Kiev 1975. 25—30.

[121] LOVE, A.: Mathematical theory of elasticity. Russian translation. AN SSSR, Moscow-Leningrad 1955.

[122] MAGNITSKY, V. A.: Internal structure and physics of the Earth (in Russian). Nedra, Moscow 1965.

[123] URAZOV, B. M.; VOLAROVICH, M. P.; KURSKEYEV, A. K.: Physical properties of rocks under deep thermodynamic conditions (in Russian). Nauka, Kaz. SSR, Alma-Ata 1973.

[124] VOLAROVICH, M. P.; GUREVICH, A. S.: Studies of the dynamic modulus of elasticity of rocks against temperature (in Russian). Izv. AN SSSR, Ser. geofiz. **4** (1957) 417—425.

[125] VOLAROVICH, M. P.: High pressure studies of elastic properties of rocks as related to the Earth's crust and upper mantle investigations (in Russian). In: Physico-mechanical properties of rocks of the upper crust. Nauka, Moscow 1968, 30—36.

[126] YAMSCHIKOV, V. S.: The modulus of elasticity against temperature studies (in Russian). Izv. AN SSSR, Ser. Fiz. Zemli **10** (1965) 83—87.

[127] VOLAROVICH, M. P.: The apparatus and experimental procedure in studies of physical properties of rocks at high pressure (in Russian). In: Unikalnye pribory, **5** (1970) 3—8.

[128] BAYUK, E. I.; VOLAROVICH, M. P.; KLIMA, K.; PROS, Z.; WANEK, J.: Velocity of longitudinal waves in eclogite and ultrabasic rocks under pressures to 4 kilobars. Studia geoph. et geod. **11** (1967) 271—279.

[129] VOLAROVICH, M. P.: Physicomechanical properties of rocks under different pressures (in Russian). In: All-union symposium of tectonophysics. Gosgeoltechizdat, Moscow 1960, 9—36.

[130] PARKHOMENKO, E. I.: Electrization phenomena in rocks (in Russian). Nauka, Moscow 1968.

[131] TOMASHEVSKAYA, I. S.; KIREENKOVA, S. M.: The apparatus for studying the effect of strain rates on strain and strength of rocks under high pressure (in Russian). In: Physical properties of rocks at high pressures and temperatures. Metsniereba, Tbilissi 1974, 286—288.

[132] TOMASHEVSKAYA, I. S.; KHAMIDULLIN, J. N.: The application of the kinetic concept of strength to rocks under triaxial compression (in Russian). In: Physical properties of rocks at high thermodynamic parameters. Naukova Dumka, Kiev 1971, 208—211.

[133] Budnikov, V. A.; Volarovich, M. P.; Levykin, A. I.: Compressional velocities in dry and water-saturated marble specimens at plastic flow under non-uniform compression. (in Russian). Geofiz. sb. AN Ukr. SSR (1975) 66, 73—78.

[134] Budnikov, V. A.: The apparatus for experiments on rock samples under complex strain (in Russian). Izv. VUZ, ser. geol. i razvedka **6** (1975) 124—128.

[135] Volarovich, M. P.; Tomashevskaya, I. S.; Khamidullin, J. N.: Study of rheologic processes and failure of rock samples under pressure to 10 kbar (in Russian). Geofiz. sb. AN Ukr. SSR (1969) 53, 40—43.

[136] Volarovich, M. P.; Bayuk, J. S.; Valius, V. P.; Galkin, I. N.: Seismic crustal models from data of explosion seismology and experimental velocity measurements (in Russian). Izv. AN SSSR, Fiz. Zemli (1975) 5, 62—67.

[137] Aslanian, A. T.; Volarovich, M. P.; Levykin, A. I.; Arutiunian, A. V.: The composition, structure and elastic properties of the crust and upper mantle in Armenia (in Russian). Dokl. AN. Arm. SSR (1975) 3, 55—59.

[138] Volarovich, M. P.; Levykin, A. I.; Elevich, V. J.: Tectonic conditions of serpentinization of ultrabasic rocks in Voronezh crystalline massif from experimental measurements of velocities in rocks under pressure to 25 kbar (in Russian). Izv. AN SSSR, Ser. geol. (1972) 12, 78—88.

[139] Zviagintsev, L. I.; Podolsky, A. M.: Petrophysical evidence of the conditions of tin-bearing granite emplacement in the Okhotsk-Chukotka volcanic belt (in Russian). Izv. AN SSSR, Ser. geol. (1975) 9, 78—93.

[140] . . . Intrusions, hydrothermal-metasomatic formations and copper-molybdenum mineralization (in Russian). Nauka, Moscow 1975.

[141] Tomashevskaya, I. S.; Khamidullin, Ya. N.: Premonitory symptoms of breakdown of rock samples (in Russian). Izv. AN SSSR, Fiz. Zemli (1972) 5, 12—20.

[142] Sih, G. C.: Fracture, ed. H. Libowitz. Academic Press, N.Y. **2** (1968).

[143] Muskhelishvili, N. I.: Some basic problems of mathematical theory of elasticity, (in Russian). Moscow 1966.

[144] Niewiadomski, J.: Analysis of crack stresses and its application to problems of orogen mechanics. Materialy i Prace, Publication of the Institute of Geophysics, Polish Academy of Sciences **85** (1975).

[145] Korneichuk, A. A.: Quadrature formulas for singular integrals (in Russian). In: Numerical solution methody of differential equations and quadrature formulas. Moscow 1967.

[146] Dmowska, R.; Kostrov, B. V.: A shearing crack in a semispace under strain conditions. Archives of Mechanics, Warsaw, **25** (1973) 3.

[147] Fung, Y. C.: Foundation of solid mechanics. Prentice-Hall, inc., Englewood Cliffs, New Jersey, 1965.

[148] Parkhomenko, E. I.; Bondarenko, A. I.: Electrical conductivity of rocks at high pressures and temperatures (in Russian). Nauka, Moscow 1972.

[149] Bradley, R. S.; Jamil, A. K.; Munro, D. C.: The electrical conductivity at high temperatures and pressures. Geochem. et Cosm. Acta **28** (1964) 1660—1678.

[150] Duba, A.; Heard, H. C.; Schock, R. N.: Electrical conductivity of olivine at high pressure and under controlled oxygen fugacity. J. geophys. Res. **79** (1974) 1667—1673.

[151] Khitarov, N. I.; Slutskij, A. B.; Revin, N. I.: Electrical conductivity of olivins at high *PT*-conditions (in Russian). In: Physical properties of rocks at high pressures and temperatures (in Russian). Metsniereba, Tbilissi 1974, 143—144.

[152] Khitarov, N. I.; Slutsky, A. B.; Revin, N. I.: Variation of the electrical conductivity of basalts during melting under high pressure conditions (in Russian). In: Physical properties of rocks at high thermodynamical parameters. Naukova Dumka, Kiev 1971, 99—100.

[153] Slutsky, A. B.: Electrical conductivity variation at cyanite-sillimanite transformation under the conditions of high pressures and temperatures (in Russian). Dokl. AN SSSR **179** (1968).

[154] Parkhomenko, E. I.; Mkrtchan, C. A.: Electrical conductivity of jadeite at phase transformations (in Russian). In: Physical properties of rocks at high pressures and temperatures. Metsniereba, Tbilissi 1974, 137—139.

[155] AKIMOTO, S.; FUJISAWA, H.: Demonstration of electrical conductivity jump produced by the olivine-spinel transition. J. geophys. Res. **70** (1965) 443—449.

[156] PARKHOMENKO, E. I.; MKRTCHAN, C. A.: Investigation of the process of the siderite, magnesite and aragonite decarbonatization by the method of electroresistivity at high pressures (in Russian). Geochimia **6** (1975) 829—834.

[157] VERESHCHAGIN, L. F.; JAKOVLEV, E. N.; VINOGRADOV, B. V.; SAKUN, B. P.: Transformations of Al_2O_3 and NaCl to conductive state (in Russian). Letters in JETP **20** (1974) 7.

[158] VERESHCHAGIN, L. F.; JAKOVLEV, E. N.; VINOGRADOV, B. V. and others: Transformation of SiO_2 to conductive state (in Russian). Letters in JETP **20** (1974) 7, 472—474.

[159] VERESHCHAGIN, L. F.; JAKOVLEV, E. N.; VINOGRADOV, B. V. and others: At transformation of diamond to conductive state (in Russian). Letters in JETP **17** (1973) 8.

[160] BOTT, M.: The interior of the Earth. Edward Arnold, London 1971.

[161] GELADZE, G. G.: On the peculiarities of determination of electric properties of rocks in the alternating fields within 100—1000 °C (in Russian). Geofiz. sb. AN Ukr. SSR (1975) 66, 89—92.

[162] ZUBOV, V. G.; FIRSOVA, M. N.; MOLOKOVA, T. M.: Temperature dependence of dielectric permeability of crystalline and melted quartz (in Russian). Kristallografia **8** (1963).

[163] BONDARENKO, A. T.: The study of temperature dependence of dielectric permeability of rocks with various frequencies (in Russian). Izv. AN SSSR, Ser. geofiz. (1963) 3, 455—463.

[164] BONDARENKO, A. T.: The effect of the dehydratation process on the electric conductivity of the metamorphic complex of rocks (in Russian). Dokl. AN SSSR **208** (1973).

[165] PARKHOMENKO, E. I.; BONDARENKO, A. T.: The effect of the dehydratation process on the set of the physical parameters of rocks with high pressures and temperatures (in Russian). In: Physics of earthquake focus. Nauka, Moscow 1975, 140—151.

[166] DYSON, F. et al.: Stability and phase transitions (Russian translation). Mir, Moscow 1970.

[167] STANLEY, T.: Phase transitions and critical phenomena (Russian translation). Mir, Moscow 1974.

[168] SEMENCHENKO, V. K.: Selected chapters of theoretical physics (in Russian). Prosveshchenie, Moscow 1966.

[169] WANG, H.; TODD, T.; WEIDNER, D.; SIMMONS, G.: Elastic properties of Apollo 12 rocks. Proc. Sec. Lunar Sci. Conf., MIT Press, **3** (1971) 2327—36.

[170] TODD, T.; WANG, H.; BALDRIDGE, W. SC.; SIMMONS, G.: Elastic properties of Apollo 14 and 15 rocks. Proc. Third Lunar Sci. Conf., MIT Press, **3** (1972) 2577—86.

[171] KANAMORI, H.; NUR, A.; CHUNG, D.; SIMMONS, G.: Elastic wave velocities of lunar samples at high pressures and their geophysical implications. Proc. Apollo 11 Lunar Sci. Conf., Pergamon Press, **3** (1970) 2286—2293.

[171a] ENGLER, R.: Elastische und thermische Eigenschaften von Mondgestein und oberen Schichten des Mondes. Diplomarbeit, Humboldt-Universität, Berlin 1976 (unpublished).

[172] SCHRAGE, I.; SCHUPPE, F.: Die Beeinflussung der dynamischen Poissonkonstante durch Porosität und Eigenschaften der Porenfüllung. Bergakademie **20** (1968) 12, 708—713.

[173] STEPHENS, D. R.; LILLEY, S.: Compressibilities of lunar crystalline rocks, microbreccia and fines to 40 kilobars. Science **167** (1970) 731—732.

[174] SCHREIBER, E.; ANDERSON, O. L.; SOGA, M.; WARREN, M.; SCHOLZ, C.: Sound velocity and compressibility for lunar rocks 17 and 46 and for glas spheres from the lunar soil. Science **167** (1970) 732—734.

[175] BALDRIDGE, W.; SIMMONS, G.: Thermal expansion of lunar rocks. Second Lunar Sci. Conf., MIT Press, **3** (1971) 2317—2321.

[176] VAN FALKENBURG, A.; WEIR, C. E.; LIPPINCOTT, E.: Modern very high pressures techniques. London 1962.

[177] PARKHOMENKO, E. J.: Electrical properties of rocks. New York 1967.

[178] OESBERG, R.-P.: Der elektrische Widerstand von Mineralen und Gesteinen unter hohen Drücken und Temperaturen und der Bau der Erdkruste und des oberen Mantels. Diss. Humboldt-Universität Berlin 1969.

[179] Strangway, D. W.; Chapman, W. B.; Olhoeft, G. R.; Carnes, J.: Electrical properties of lunar soil dependence on frequency, temperature and moisture. Earth and planet. Sci. Lett. **16** (1972), 275—281.

[180] Schwerer, F. C.; Huffmann, G. P.; Fisher, R. M.: Electric conductivity of lunar surface rocks at elevated temperatures. Proc. 4th Lunar Sci. Conf. Houston (Texas) 1974.

[181] Berdichevsky, M. N. et al.: Conductivity layers in the Earth's crust and upper mantle. Gerlands Beitr. z. Geophysik **81** (1975) 187—196.

[182] Chan, T.; Nyland, E.; Gough, D.: Partial melting and conductivity anomalies in the upper mantle. Nature phys. science **244** (1973) 89—91.

[183] O'Konsky, C. T.: Electrical properties of macromolecules. J. phys. chemistry (1960) 5.

[184] Chelidze, T. L.: On the effect of surface polarization upon electrical properties of rocks in alternating fields (in Russian). Izv. AN SSSR, Fiz. Zemli (1969) 10. 80.

[185] Chelidze, T. L.: Electric conductivity and dielectric permeability of rocks (in Russian). In: Rocks in physical fields. Metsniereba, Tbilissi 1971, 53—109.

[186] Hanai, T.: Electrical Properties of Emulsions. In: Emulsion Science, ed. by P. Scherman. Academic Press 1970.

[187] Geladze, G. G.; Chelidze, T. L.: Frequencies dependency of specifical electric resistivity and dielectric permeability on magnetite — bearing rocks at high temperatures (in Russian). In: Physical properties of rocks at high thermodynamic parameters. Naukova Dumka, Kiev 1971, 69—72.

[188] Dukhin, S. S.; Sorokina, T. S.; Chelidze, T. L.: On the theory of low-frequency dispersion of dielectric permeability of the moistened disperse System (in Russian). Colloidal J. **31** (1969) 823—828.

[189] Chelidze, T. L.: On the definition of surface conductivity. Elektronnaya obrabotka metallov, Kishinew, (1973) 4, 73.

[190] Schwarz, G.: A theory of the low-frequency dielectric dispersion of colloidal particles in electrolytic solutions. J. chem. Phys. **66** (1962) 2636.

[191] Dukhin, S. S.; Shilov, V. N.: Dielectric phenomena and a double layer in disperse systems and polyelectrolytes (in Russian). Naukova Dumka, Kiev 1972.

[192] Chelidze, T. L.: Surface effects in dielectric spectroscopy of heterogenous systems (in Russian). Doct. Thesis, Kiev 1974.

[193] Matare, H.: Defect electronics in semiconductors. Wiley-Interscience, New York, London 1971.

[194] Chelidze, T. L.; Geladze, G. G.: On the mechanism of anomalous frequency dispersion of rocks at high temperatures (in Russian). In: Physical properties of rocks at high pressures and temperatures. Metsniereba, Tbilissi 1974, 125—127.

[195] Chelidze, T. L.; Shilov, V. N.: Determination of the potential's full jump in the diffuse double layer by dielectrical method (in Russian). Colloidal J. **35** (1973) 105—109.

[196] Chelidze, T. L.: On the effect of surface physical and chemical factors on the dielectrical properties of the disperse systems (in Russian). In: Electric surface phenomena in disperse systems. Nauka, Moscow 1972, 40—45.

[197] Yund, R. A.; Kullerud, G.: The Ni-S-system and related minerals. J. Petrology **3** (1962) 126—175.

[198] Ward, J. C.: The structure and properties of some iron sulphides. Rev. Pure and Appl. Chem. **20** (1970) 175—206.

[199] Fleet, M. E.: On the lattice parameters and superstructures of pyrrhotites. Amer. Mineralogist **53** (1968) 1846—1855.

[200] Morimoto, N.; Gyobu, A.; Mukaiyama, H.; Izawa, E.: Crystallography and stability of pyrrhotites. Econ. Geology **70** (1975) 824—833.

[201] Naldrett, A. J.; Craig, J. R.; Kullerud, G.: The central portion of the Fe–Ni–S system and its bearing on pentlandite exsolution in iron-nickel sulfide ores. Econ. Geology **62** (1967) 826—847.

[202] Shewman, R. W.; Clark, L. A.: Pentlandite phase relations in the Fe–Ni–S system and notes on the monosulfide solid solution. Can. J. of Earlth Sciences **7** (1970) 67—85.

[203] Kullerud, G.: The Fe–Ni–S system. Carnegie Instit. Washington, Yearbook **62** (1963) 175—189.

[204] Kullerud, G.: Thermal stability of pentlandite. Canad. Mineral. **7** (1963) 353—366.

[205] Morimoto, N.; Gyobu, A.; Tsukuma, K.; Koto, K.: Superstructure and nonstoichiometry of intermediate pyrrhotites. Amer. Miner. **60** (1975) 240—248.

[206] MISRA, K. C.; FLEET, M. E.: Unit cell parameters of monosulfide, pentlandite and taenite solid solutions within the Fe–Ni–S system. Mat. Res. Bull. **8** (1973) 669—678.

[207] KRABBES, G.; MATTERN, N.: Chemischer Transport von Eisensulfid (IV). Die Charakterisierung des Eisensulfids mittels röntgenographischer Methoden. Kristall und Technik **11** (1976) 767—774.

[208] ARNOLD, R. G.; REICHEN, L. E.: Measurement of the metal content of naturally occuring metal-deficient hexagonal pyrrhotite by an X-ray spacing method. Amer. Mineral. **47** (1962) 105—111.

[209] BEZUGLAYA, L. S.; MARTYNOVA, T. A.; PETROVA, G. N.; RYBAK, R. S.: Ascertaining the origin of the rock magnetization by method of comparative characteristics of stability on an example of iron cherts from KMA (in Russian). Izv. AN SSSR, Ser. geofiz. **4** (1962) 514—523.

[210] ... Geology and genesis of the Precambrian iron chert and manganese formations of the world (Proc. of Intern. Symp.) (in Russian). Naukova Dumka, Kiev 1972.

[211] GRABOVSKY, M. A.; PARKHOMENKO, E. I.: On the change of the magnetic properties of magnetite under the effect of high releasing stresses (in Russian). Izv. AN SSSR, Ser. geofiz. **5** (1953) 405—417.

[212] KOPAY'EV, V. V.; MARTYNOVA, T. A.: The experience of application of the laboratory measurements of quartzites to interpretation of magnetic anomalies of KMA (in Russian). Izv. AN SSSR, Ser. geofiz. **4** (1961) 553—566.

[213] KRUTIKHOVSKAYA, Z. A.; ZAVOYSKY, V. N.; PODOLYANKO, S. M.; SAVENKO, B. YA.: Magnetization of rocks from the iron-ore formations of the Great Krivoy Rog and KMA (in Russian). Naukova Dumka, Kiev 1964.

[214] KRUTIKHOVSKAYA, Z. A. et al.: The structure of the basement and iron-ore deposits of the northern slope of the Ukrainian Shield (in Russian). Naukova Dumka, Kiev 1975.

[215] KUGHELOV, K. K.; KRUTIKHOVSKAYA, Z. A.: Formation of remanent magnetization and its distribution within rocks (in Russian). Sov. geol. **2** (1960) 125—139.

[216] KUTAS, R. I.; GORDIENKO, V. V.: The thermal field of the Ukraine (in Russian). Naukova Dumka, Kiev 1971.

[217] LEBEDEV, T. S.; POZNANSKAYA, N. F.: Some problems of the theoretical modelling of the behaviour of magnetic susceptibility and remanent magnetization of rocks on the condition of high pressure and temperature (in Russian). In: Physical properties of rocks under high thermodynamic parameters. Naukova Dumka, Kiev 1971, 75—77.

[218] LEBEDEV, T. S.; POZNANSKAYA, N. F.: Modelling of the changes in magnetic properties of magnetite under the deep thermodynamic conditions of the Ukrainian Shield (in Russian). Geofiz. sb. AN Ukr. SSR (1972) 57, 33—36.

[219] LEBEDEV, T. S.; POZNANSKAYA, N. F.: Theoretische Modellierung von Veränderungen magnetischer Eigenschaften eines mineralischen Materials bei hohen Drücken und Temperaturen. Gerlands Beiträge z. Geophysik **84** (1975) 335—347.

[220] LEBEDEV, T. S.; SAVENKO, B. YA.; INDUTNY, V. F.: The behaviour of remanent magnetization of the basement rocks of the Dnieper-Donets Depression under high hydrostatic pressure (in Russian). Geofiz. sb. AN Ukr. SSR (1974) 61, 89—96.

[221] LEBEDEV, T. S.; SAVENKO, B. YA.; SHAPOVAL, V. I.; KORCHIN, V. A.: A device for the study of magnetic properties of rocks under hydrostatic pressure to 2,5 kbar (in Russian). Geofiz. sb. AN Ukr. SSR (1972) 46, 85—87.

[222] MARTYNOVA, T. A.: Certain regularities of the magnetic characteristics behaviour dependent on the conditions of formation and life of the rock on an example of iron cherts from KMA (in Russian). In: The present and past magnetic field of the Earth. Nauka, Moscow 1965, 147—153.

[223] MELNIK, Y. P.: Thermodynamic constants for the analysis of the conditions of the iron ores formation (in Russian). Naukova Dumka, Kiev 1972.

[224] MELNIK, Y. P.: Physico-chemical conditions of the iron chert formations in the Precambrian (in Russian). Naukova Dumka, Kiev 1973.

[225] SOLLOGUB, V. B.; CHEKUNOV, A. V. et al.: The main results and problems of the deep crustal structure studies in the Ukraine by the explosion seismology methods (in Russian). Geofiz. sb. AN Ukr. SSR (1970) 38, 48—63.

[226] Stebnovskaya, Y. M.; Yaroschuk, M. A.; Ivanitsky, V. I.: Genetical properties of magnetite in iron cherts from the data of the nuclear gamma-resonance and thermomagnetic analyses (in Russian). In: Mineralogy of sedimentary formations. Naukova Dumka, Kiev, **3** (1976) 33—41.

[227] Dunlop, D. J.: Thermal enhancement of magnetic susceptibility. J. Geophys. **40** (1974) 439—451.

[228] Radhakrishnamurty, C.; Likhite, S. D.: Hopkinson effect, blocking temperature and Curie point in basalts. Earth Planet. Sci. Letters **7** (1970) 389—396.

[229] Radhakrishnamurty, C.; Deutsch, E. R.: Magnetic techniques for ascertaining the nature of iron oxide grains in basalts. J. Geophys. **40** (1974) 453—465.

[230] Träuble, H.: Moderne Probleme der Metallphysik. Springer-Verlag 1966, 420—459.

[231] Jelenska, M. B.: The effect of compression on the magnetic properties of rocks. Materialy in Prace, Publication of the Institut of Geophysics, Polish Academy of Sciences **72** (1974).

[232] Parker, W. J. et al.: Flash method of determining thermal diffusivity, heat capacity, and thermal conductivity. J. Appl. Physics **32** (1961) 9, 1679—1684.

[233] Dzhavadov, L. N.: Measurement of the specific heat of metals under pressure. Specific heat of the high-pressure phases of bismuth. High Temper. — High Press. **5** (1973) 4, 455—458.

[234] Loriers-Susse, C.; Bastide, J.-P.; Baeckstroem, G.: Specific heat measured at high pressures by a pulse method. Rev. Scient. Instr. **44** (1973) 9, 1344—1349.

[235] Taylor, R. E.; Clark III, L. M.: Finite pulse time effects in flash diffusivity method. High Temper. — High Press. **6** (1974) 65—72.

[236] Lincoln, R. C.; Donaldson, A. B.; Heckman, R. C.: High-temperature thermal-diffusivity measurement by a negative-pulse technique. J. Appl. Physics **45** (1974) 5, 2321—2326.

[237] Carslaw, H. S.; Jaeger, J. C.: Conduction of heats in solids. Oxford, Clarendon Press 1959, second edition.

[238] Lubimova, E. A.; Starikova, G. N.; Shushpanova, A. P.: Thermophysical investigations of rocks (in Russian). In: Geothermal Investigations. Nauka, Moscow 1964.

[239] Tykhonov, A. N.; Samarski, A. A.: Differentialgleichungen in der Mathematischen Physik. VEB Deutscher Verlag Wissensch., Berlin 1959.

[240] Seipold, U.; Gutzeit, W.: Untersuchungen der Druckabhängigkeit der Wärmeleitfähigkeit einiger Gesteine. Gerlands Beitr. z. Geophysik **83** (1974) 6, 498—504.

[241] Kondratenko, A. M.; Nersesov, I. L.: Some results of the study on changes in the velocity of longitudinal and transversal waves in a focal zone (in Russian). Izv. AN SSSR, Fiz. Zemli **25** (1962) 130—150.

[242] Semenov, A. N.: Change of longitudinal and shear wave travel time ratio before strong earthquakes (in Russian). Izv. AN SSSR, Fiz. Zemli (1969) 4, 72—77.

[243] Nersesov, I. L.; Semenov, A. N.; Simbireva, I. G.: Areal distribution of longitudinal and shear wave travel time ratio about Garm-district (in Russian). In: Physical basis of earthquake forerunners. Nauka, Moscow 1969.

[244] Aggarwal, Y. P.; Sykes, L. R.; Armbruster, J.; Sbar, M. L.: Premonitory changes in seismic velocities and predictions of earthquakes. Nature **241** (1973) 101—104.

[245] Whitcomb, J. H.; Garmany, J. H.; Anderson, D. L.: Earthquake prediction: Variation of seismic velocities before San Fernando earthquake. Science **180** (1973) 632—635.

[246] Brady, B. T.: Theory of earthquakes. Pure and appl. Geophysics **112** (1974) 701—727.

[247] Nur, A.: Dilatancy, pore fluids and premonitory variations of the t_s/t_p travel times. Bull. Seism. Am. **62** (1972) 1217—1222.

[248] Scholz, C. H.; Sykes, L. R.; Aggarwal, Y. R.: Earthquake prediction: A physical basis. Science **181** (1973) 803—810.

[249] Mjachkin, V. I.; Brace, W. F.; Sobolev, G. A.; Dietrich, J. W.: Two models for earthquake forerunners. Pure and appl. Geophysics **113** (1975) 169—182.

[250] Walsh, J. B.: The effect of cracks on the compressibility of rocks. J. geophys. Res. **70** (1965) 381—389.

[251] Walsh, J. B.: The effect of cracks on the uniaxial elastic compression of rocks. J. geophys. Res. **70** (1965) 399—411.

[252] Walsh, J. B.: The effect of cracks in rocks on the POISSON ratio. J. geophys. Res. **70** (1965) 5249 to 5257.

[253] Stiller, H.; Wagner, F. C.; Heinrich, R.: Der Einfluß der Porengröße und des Druckes auf die Schallgeschwindigkeit poröser Medien. Neue Bergbautechnik **2** (1972) 326—330.

[254] Schön, J.: Vergleichende Betrachtungen zur Druckabhängigkeit der Ausbreitungsgeschwindigkeit elastischer Wellen an lunaren und terrestrischen Gesteinsproben. Gerlands Beitr. z. Geophys. **82** (1973) 472—476.

[255] Stiller, H.; Wagner, F. C.; Freund, D.; Voigt, M.: Eine Methode zur Bestimmung der Schallgeschwindigkeit v_p poröser, flüssigkeitsgesättigter Gesteine unter Gebirgsdruck von 2000 at. Neue Bergbautechnik **1** (1971) 333—339.

[256] Nagumo, S.: Compaction of sedimentary rock. A consideration by the theory of porous media. Bull. Earthq. Res. Inst., Tokyo, **43** (1965) 339—348.

[257] Stiller, H.; Wagner, F. C.; Vollstädt, H.: A model for the v_p-pressure function of cracked lunar rocks and some possible seismological consequences. Tectonophysics **31** (1976) 129—137.

[258] Stiller, H.; Wagner, F. C.; Vollstädt, H.: The dependence of longitudinal wave velocity on pressure in cracked rocks. XIVth General Assembly of the European Seismological Commission, Trieste, 16—22 September 1974, Nationalkomitee für Geödäsie und Geophysik bei der Akademie der Wissenschaften der Deutschen Demokratischen Republik, Berlin 1975, 199—202.

[259] . . . Electrical and mechanical properties of rocks under high pressure. Izv. AN SSSR, Fiz. Zemli **37** (1966).

[260] Wagner, F. C.; Voigt, M.: Eine Methode zur Bestimmung der Porosität und Kompressibilität poröser Gesteine unter Gebirgsdruck. Z. angew. Geologie **17** (1971) 413—416.

[261] Stiller, H.; Wagner, F. C.; Vollstädt, H.: A 2-phase-model for description of the influence from cracks on the v_p and v_s wave-velocities in dry and saturated rocks. Tectonophysics (in press) 1976.

[262] Fisz, M.: Wahrscheinlichkeitsrechnung und mathematische Statistik. Berlin 1958.

[263] Todd, T.; Simmons, G.: Effect of pore pressure on the velocity of compressional waves in low-porosity rocks. J. geophys. Res. **77** (1972) 3731—3743.

[264] Azbel, I. Ya.; Keilis-Borok, V. I.; Yanovskaya, T. B.: A technique for the joint interpretation of travel-time curves and amplitude curves in studying the upper mantle (in Russian). Numerical seismology, Nauka, **2** (1966).

[265] Valus, V. P.: Determination of seismic cross-section from combined data (in Russian). Numerical seismology **4** (1968).

[266] Petrashen, G. I.; Alekseev, A. S.; Gel'chinsky, B. Ya.: Elements of the dynamic theory of the propagation of elastic waves (in Russian). In: Some questions about the dynamical theory of seismic waves propagation, **3** (1959).

[267] Valus, V. P.; Levshin, A. L.; Sabitova, T. M.: Joint interpretation of body and surface waves for a region in Central Asia (in Russian). Numerical seismology **2** (1966).

[268] Valus, V. P.; Veitsman, P. S.; Galkin, I. N.; Levshin, A. L.: On non-uniqueness of the velocity cross-section of the crust in DSS (in Russian). Alma Ata 1969.

[269] Valus, V. P.; Galkin, I. N.; Levshin, A. L.: Computer solution of the inverse problem of DSS from a travel-time curve and an amplitude curve (in Russian). Numerical seismology **5** (1971).

[270] Galkin, I. N.: Determining the seismic model of the Earth's crust (in Russian). Nauka, Moscow 1972.

[271] Volarovich, M. P.: Studies in the physical properties of rocks at high pressures and temperatures (in Russian). Geofiz. sb. AN Ukr. SSR (1964) 9, 3—13.

[272] Afanasyev, G. D.; Bayuk, E. I.; Belikov, B. R.; Volarovich, M. P.; Zalessky, B. V.; Pavlogradsky, V. A.; Sinyakov, I. Z.: Physical properties of rocks from the Northern Caucasus (in Russian). In: Physico-mechanical properties of rocks from the upper part of the Earth's crust. Nauka, Moscow 1968, 8—30.

[273] Bayuk, E. I.; Volarovich, M. P.; Skvortsova, L. S.: The velocities of elastic waves at high pressures in igneous and metamorphic rocks (in Russian). In: Tectonophysics and mechanic properties of rocks. Nauka, Moscow 1971, 127—137.

[274] Volarovich, M. P.; Levykin, A. I.; Skorikova, M. F.: The elastic properties of rocks from Primorye Territory under pressures up to 20000 kg/cm^2 (in Russian). Trudy Sakhknii **24** (1970).

[275] VOLAROVICH, M. P.; KURSKEEV, A. K.; TOMASHEVSKAYA, I. S.; TUZOVA, I. L.; URAZAEV, B. M.: A study on the velocity of *P*-wave in rock specimens at high pressures in connection with investigation of the deep structure of Central Kazakhstan (in Russian). In: Tectonophysics and mechanic properties of rocks. Nauka, Moscow 1971, 151—161.

[276] SKORIKOVA, M. F.: The elastic properties of rocks from Primorye territory under pressures up to 20000 kg/cm² (in Russian). Nauka, Moscow 1970.

[277] BAYUK, E. I.; VOLAROVICH, M. P.; GALDIN, N. E.: High pressure study of the velocities of elastic waves in rock samples from the rift zone and islands of the Indian ocean. In: Studies on the problem of the rift zones of the world ocean. Nauka, 1972.

[278] LATHAM, G. et al.: Lunar structure and dynamics-results from the Apollo passive seismic experiment. The Moon **7** (1973) 544.

[279] LACAM, A.; PEYRONNEAU, J.: Transition polymorphique de RbCl sous haute pression. Influence des paramètres expèrimentaux sur léffet dhystérésis. J. Phys. **34** (1973) 1047—1053.

[280] LACAM, A.; PEYRONNEAU, J.: Transitions polymorphiques ultra-rapides induites par la pression, dans les halogénures alcalins. Rev. Phys. Appl. **10** (1975) 299—304.

[281] LELIWA KOPYSTYNSKI, J.; PEYRONNEAU, J.; LACAM, A.: Étude cinétique de léffet dhystérésis dans les transitions polymorphiques. Influence de la surpression. J. Phys. **35** (1974) 609—614.

[282] LELIWA KOPYSTYNSKI, J.: Kinetics of sudden polymorphic transformations occurring under high pressure. Publications Inst. Geoph. Pol. Ac. Sci. **101** (1976) 133—143.

[283] LELIWA KOPYSTYNSKI, J.; PEYRONNEAU, J.; LACAM, A.: Successive sudden high pressure polymorphic transformations-Experiments — Mathematical model — Geophysical implications. (in press).

[284] WILKE, K.-TH.: Methoden der Kristallzüchtung. Dt. Verlag d. Wissenschaften, Berlin 1975.

[285] FRÖLICH, F.: Beiträge zum Erkundungsprogramm: Materieparameter im Bereich der Erdkruste. Teil I: Verfahrensgrundlagen, Erprobungsergebnisse, Testtechnik. Geomagn. Inst. Potsdam, Abh. 41 (1969).

[286] FRÖLICH, F.: Teil II: Ergänzende festkörperphysikalische und physiko-chemische Untersuchungen, Auswertung. ZI Physik der Erde, Potsdam, Veröff. Nr. 4 (1970).

[287] TREDER, H.-J.: Kosmogonie und Erdentstehung. In: Physik des Planeten Erde, ed. by R. LAUTERBACH. Akademie-Verlag, Berlin 1975.

[288] UMC Monograph: The Earth's crust and upper mantle. UMP Scientific Report No. 21, Am. Geophys. Union, (1969) 1—726.

[289] RINGWOOD, A. E.; MAJOR, A.: Synthesis of Mg_2SiO_4-Fe_2SiO_4 solid solutions. Earth plan. sci. letters **1** (1966) 241.

[290] GRAHAM, E. K.: On the compression of stishovite. Geophys. J. R. Astr. Soc. **32** (1973) 15—34.

[291] HINZ, W.: Silikate (Bd. I und II). Verlag f. Bauwesen, Berlin 1970.

[292] FRÖLICH, F.: Tiefenbedingte Gesetzmäßigkeiten des Gesteinszustandes im Erdinneren. Gerlands Beitr. Geophys. **82** (1973) 397—405.

[293] COOK, A. H.: The dynamical properties and internal constitution of the Earth, the Moon, and the planets. Proc. Roy. Soc. London, A **328** (1972) 301—336.

[294] FRÖLICH, F.: Neue Aspekte zum GEODYNAMICS PROJECT aus hochdruck-festkörperphysikalischer und planetarer Sicht. ZI Physik der Erde. Veröff. Nr. 29 (1974).

[295] HIGGINS, G.; KENNEDY, G. C.: The adiabatic gradient and the melting-point gradient in the core of the Earth. J. Geophys. Res., **76** (1971) 1870—1878.

[296] BOLT, B. A.: Structure of the Earth's core from seismological evidence. Transact. AGU., **53** (1972) 599.

[297] PHINNEY, R. A.: Seismological evidence on the coremantle boundary. Transact. AGU, **53** (1972) 600.

[298] FRÖLICH, F.: Festkörperphysikalische Untersuchungen im Zusammenhang mit Ergebnissen der Hochdruck-Hochtemperatur-Physik. ZI Physik der Erde, Potsdam, Veröff. Nr. 17 (1972).

[299] FRÖLICH, F.; SEIPOLD, U.; VOLLSTÄDT, H.; WALZER, U.: Complex studies on variations of the secular behaviour in the Earth's mantle. Monographs on terrestrial, solar, and cosmic physics. Vol. I. Akad. Verl.-Ges. Geest & Portig, Leipzig 1973.

[300] KAWAI, N.; MOCHIZIKI, S.: Insulator-metal transition in NiO. Phys. Letters, A **36** (1971) 54—55.

[301] CLARK, S. P.: Radiative transfer in the Earth's mantle. Transact. AGU, **38** (1957) 931ff.

[302] v. Herzen, R. P.: Surface heat flow and some implications for the mantle. In: The Earth's mantle. Acad. Press, London—New York 1967.

[303] Lubimova, H. A.: Thermal history of the Earth — with consideration of the variable thermal conductivity of its mantle. Geophys. J. **1** (1958) 195ff.

[304] Lubimova, H. A.: Theory of thermal state of the Earth's mantle. In: The Earth's mantle, Acad. Press, London—New York 1967.

[305] Schmucker, U.: Geophysical aspects of structure and composition of the Earth. In: Hand book of Geochemistry, Vol I. Springer-Verlag, Berlin—Heidelberg—New York 1969.

[306] Walzer, U.: Übersicht zu Arbeiten über Konvektion im Erdmantel. ZI Physik d. Erde, Potsdam, Veröff. Nr. 21 (1973).

[307] Walzer, U.: A convection mechanism for explaining episodicity of magmatism and orogeny. PAGEOPH, **112** (1974/I) 106—117.

[308] Frölich, F.: Komplexuntersuchungen an Prozeßmodellen zur Paläogeodynamik. ZI Physik d. Erde, Potsdam, Veröff. Nr. 21 (1973).

[309] Schmidt, O. Yu.: On the origin of visual binary stars and the character of their orbits (in Russian). Dokl. AN SSSR **44** (1944) 8—12.

[310] Schmidt, O. Yu.: A meteoric theory of the origin of the Earth and the planets (in Russian). Dokl. AN SSSR **45** (1944) 229—233.

[311] Schmidt, O. Yu.: A theory of the origin of the Earth. (4 lectures) Lawrence & Wishart, London 1959.

[312] Schmidt, O. Yu.: On the law of planetary distances (in Russian). Dokl. AN SSSR **52** (1946) 667—672.

[313] Urey, H. C.: The planets — their origin and development. Yale-Univeristy Press 1952.

[314] Treder, H.-J.: Thesen zur Kosmogonie und Erdentstehung. Ber. Dt. Ges. Geol. Wiss., A **16** (1971) 201—205.

[315] Treder, H.-J.: Elementare Kosmologie. Reihe WTB, Akademie-Verlag, Berlin 1975.

Author's Addresses

AKULOVA, R. S. (1)
BAYUK, E. I. (1)
BURTNY, P. A. (2)
CHELIDZE, T. L. (3)
ENGLER, R. (4)
FREUND, D. (4)
FRÖLICH, F. (4)
GALKIN, I. N. (1)
GELADZE, G. G. (3)
GUTZEIT, W. (4)
JELENSKA, M. (5)
KOPYSTYNSKI, J. M. (5)
KORCHIN, V. A. (2)
KRAFT, A. (4)
LEBEDEV, T. S. (2)
MÜLLER, H. J. (4)
NIEWIADOMSKI, J. (5)
OROVETSKY, YU. P. (2)
PARKHOMENKO, E. I. (1)
POZNANSKAYA, N. F. (2)
RAAB, S. (4)
SAVENKO, B. YA. (2)
SEIPOLD, U. (4)
SHAPOVAL, V. I. (2)
STEBNOVSKAYA, YU. M. (2)
STILLER, H. (4)
STROMEYER, D. (4)
TOMASHEVSKAYA, I. S. (1)
VALUS, V. P. (1)
VASILYAKA, V. T. (2)
VOLAROVICH, M. P. (1)
VOLLSTÄDT, H. (4)
WÄSCH, R. (4)
WAGNER, F. C. (4)
ZVIAGINTSEV, L. I. (1)

(1) Institute Physics of the Earth, Academy of Sciences of the USSR Bolshaya Gruzinskaya 10, 123810 Moscow D-242, USSR
(2) Institute of Geophysics, Academy of Sciences of the Ukr. SSR Palladina 32, 252068 Kiev — 68, USSR
(3) Institute of Geophysics, Academy of Sciences of the Georg. SSR Ruchadze Street 1, 380060 Tbilissi — 60, USSR
(4) Central Earth's Physics Institute, Academy of Sciences of the GDR Telegrafenberg, 15 Potsdam, GDR
(5) Institute of Geophysics, Polish Academy of Sciences Pasteura 3, 02-093 Warsaw, Poland

Bei Fragen zur Produktsicherheit wenden Sie sich bitte an:
If you have any questions regarding product safety,
please contact:

Walter de Gruyter GmbH
Genthiner Straße 13
10785 Berlin
productsafety@degruyterbrill.com